BELMONT UNIVERSITY LIBRARY
BELMONT UNIVERSITY
1900 BELMONT BLVD.
NASHVILLE, TN 37212

D1406460

Simple Science Experiments

AAX-7228

J,AC
D57
S67

Hans Jürgen Press

Simple Science Experiments

7715

BELMONT UNIVERSITY LIBRARY
CURRICULUM LAB

Discovery Toys

Originally published in Germany under the
title "Spiel – das Wissen schafft"
© Otto Maier Verlag Ravensburg 1964, 1967
Printed in West Germany

Contents

Astronomy
Image of the Sun

Place a pair of binoculars in an open window in the direct path of the sun's rays. Stand a mirror in front of one eyepiece so that it throws an image of the sun on to the opposite wall of the room. Adjust the mirror until the image is sharp, and darken the room.

You would risk damaging your eyes if you looked directly at the sun through binoculars, but you can view the bright disk on the wall as large and clear as in the movies. Clouds and birds passing over can also be distinguished and, if the binoculars are good, even sunspots. These are a few hot areas on the glowing sphere, some so big that many terrestrial globes could fit into them. Because of the earth's rotation, the sun's image moves quite quickly across the wall. Do not forget to re-align the binoculars from time to time onto the sun. The moon and stars cannot be observed in this way because the light coming from them is too weak.

2

Sun clock

Place a flowerpot with a long stick fixed into the hole at the bottom in a spot which is sunny all day. The stick's shadow moves along the rim of the pot as the sun moves. Each hour by the clock mark the position of the shadow on the pot. If the sun is shining, you can read off the time.

Because of the rotation of the earth the sun apparently passes over us in a semi-circle. In the morning and evening its shadow strikes the pot superficially, while at midday, around 12 o'clock, the light incidence is greatest. The shadow can be seen particularly clearly on the sloping wall of the pot.

3

A watch as a compass

Hold a watch horizontally, with the hour hand pointing directly to the sun. If you halve the distance between the hour hand and the 12 with a match, the end of the match points directly to the south.

In 24 hours the sun 'moves', because of the earth's rotation, once around the earth. But the hour hand of the watch goes twice round the dial. Therefore before midday we halve the distance from the hour hand to the 12, and after midday from the 12 to the hour hand. The match always points to the south. At midday, at 12 o'clock, the hour hand and the 12 both point to the sun standing in the south.

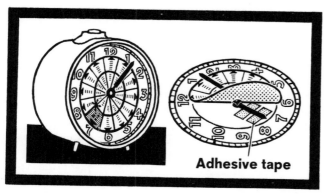

Adhesive tape

The earth rotates in 24 hours from west to east once on its axis. In this time the sun shines on all regions of the globe one after the other and determines their time of day. To enable a practical calculation of the time, the earth is divided into 24 time zones, which are very simply shown on the map below. Since in a few areas which belong together a uniform time has been introduced, the boundaries of the time zones sometimes run along state boundaries. For example, Mexico has Central time. The West European countries including Great Britain, have together with the Middle European countries, Middle-European time. According to the map, when it is 13.00 hours there it is only 7 o'clock in the morning on the East Coast of the U.S.A. in Japan it is already 21.00 hours and on the right edge (dateline) a new day is beginning.

The time zones are shown on the world time disk pictured below. Copy or stick this onto a piece of cardboard and cut it out. Colour the panel corresponding to

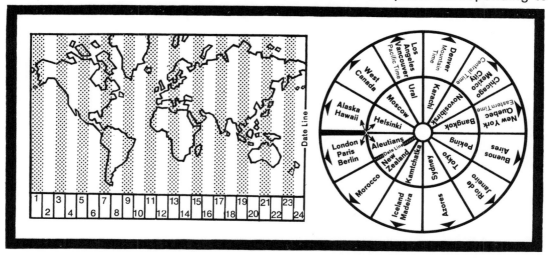

the time zone where you live red. Remove the casing and glass from an alarm clock, push the minute hand through the hole in the paper disk and fix it firmly to the hour hand. Make sure that the red-coloured panel is exactly over the hour hand. If you rotate the disk with this, it should not stick. The clock will tell you all the times of the day on the earth.

Read off first on the red panel the time of the place where you live. If you rotate the disk to the left, you will find the time zones of places west of you. In each panel, the time is an hour earlier. If you rotate to the right, you will find the places east of you. In each panel the time is an hour later. The outer circle continues into the inner circle at the crossed arrows and vice-versa. For example: in New York it is 6.15 in the morning. Then it is already 20.15 in Tokyo and in New Zealand a new day will begin in 45 minutes. Or: in London it is 20.03. What time is it in San Francisco? Look at the world map: San Francisco lies in the time zone of Los Angeles. On the rotating disk go to the left to the Los Angeles panel. The time is: 11.03.

Plant a sprouting potato in moist soil in a pot. Place it in the corner of a shoe box and cut a hole in the opposite side. Inside stick two partitions, so that a small gap is left. Close the box and place it in a window. After a couple of days the shoot has found its way through the dark maze to the light.

Plants have light-sensitive cells which guide the direction of growth. Even the minimum amount of light entering the box causes the shoot to bend. It looks quite white, because the important green colouring material, chlorophyll, necessary for healthy growth, cannot be formed in the dark.

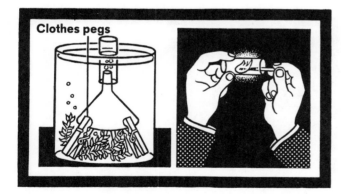

Clothes pegs

Fill a large glass jar with fresh water and place in it several shoots of waterweed. Place the jar in sunlight, and at once small gas bubbles will rise in the water. Invert a funnel over the plants and over it a water-filled glass tube. The gas which is given off by the plants slowly fills the tube.

Plants use sunlight. With its help, in the presence of chlorophyll, they make their building material, starch, from water and carbon dioxide, and give off oxygen. Oxygen has actually collected in the glass tube. If you remove the tube and hold a glowing splint in it, the splint will burn brightly.

7

Automatic watering

Fill a bottle with water and place it upside down and half buried in soil in a flower box. An air bubble rises up in the bottle from time to time, showing that the plants are using the water. The water reservoir is enough for several days, depending on the number of plants and the weather.

Water only flows from the bottle until the soil round it is soaked. It starts to flow again only when the plants have drawn so much water from the soil that it becomes dry, and air can enter the bottle. One notices that plants can take water more easily from loose soil than from hard.

8

Secret path

Dissolve a teaspoonful of salt in a glass of water and cover it tightly with parchment paper. Place the glass upside down in a dish containing water strongly coloured with vegetable dye. Although the parchment paper has no visible holes, the water in the glass is soon evenly coloured.

The tiny particles of water and dye pass through the invisible pores in the parchment paper. We call such an exchange of liquids through a permeable membrane, osmosis. All living cells are surrounded by such a membrane, and absorb water and dissolved substances in this way.

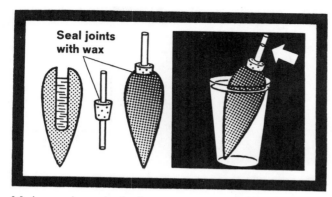

Make a deep hole in a carrot and fill it with water in which you have dissolved plenty of sugar. Close the opening firmly with a bored cork, and push a plastic straw through the hole. Mop up any overflowing sugar solution, and seal the joints with melted candle wax. Put the carrot into water and watch: after some time the sugar solution rises into the straw.

The water particles can enter the carrot through the cell walls, but the larger sugar particles cannot come out. The sugar solution becomes diluted and rises up the tube. This experiment on osmosis illustrates how plants absorb water from the soil and carry it upwards.

Fill a wine glass to overflowing with dried peas, pour in water up to the brim, and place the glass on a metal lid. The pea heap becomes slowly higher and then a clatter of falling peas begins, which goes on for hours.

This is again an osmotic process. Water penetrates into the pea cells through the skin and dissolves the nutrients in them. The pressure thus formed makes the peas swell. In the same way the water necessary for life penetrates the walls of all plant cells, stretching them. If the plant obtains no more water, its cells become flabby and it wilts.

11

Rain in a jar

Place a green twig in a glass of water in sunlight. Pour a layer of oil on to the surface of the water and invert a large jar over the lot. After a short time, drops of water collect on the walls of the jar. Since the oil is impermeable, the water must come from the leaves. In fact the water which the plant absorbs is given off into the air through tiny pores in the epidermis of the leaf. Air saturated with moisture and warmed by the sun deposits drops like fine rain on the cool glass.

12

Zig-zag growth

Lay pre-germinated seeds on a sheet of blotting paper between two panes of glass, pull rubber bands around the panes and place in a water container in a window. Turn the glass panes with the shoots onto a different edge every two days. The roots always grow downwards and the stem grows upwards.

Plants have characteristic tendencies. Their roots strive towards the middle of the earth and the shoots go in the opposite direction. On slopes the roots of trees do not grow at right-angles to the surface into the ground, but in the direction of the middle of the earth.

13

Leaf skeleton

Place a leaf on blotting paper and tap it carefully with a clothes brush, without pressing too hard or moving sideways. The leaf is perforated until only the skeleton remains, and you can see the fine network of ribs and veins.

The juicy cell tissue is driven out by the bristles and sucked up by the blotting paper. The ribs and veins consist of the firmer and slightly lignified framework and resist the brush.

14

Two Coloured Flower

Dilute red and green fountain pen inks with water and fill two glass tubes each with one colour. Split the stem of a flower with white petals, e.g. a dahlia, rose or carnation, and place one end in each tube. The fine veins of the plant soon become coloured, and after several hours the flower is half red and half blue.

The coloured liquid rises through the hair-fine channels by which the water and food are transported. The dye is stored in the petals while most of the water is again given off.

Cut a red cabbage leaf into small pieces and soak in a cup of boiling water. After half an hour pour the violet-coloured cabbage water into a glass. You can now use it for crazy colour magic. Place three glasses on the table, all apparently containing pure water. In fact only the first glass contains water, in the second is white vinegar and in the third water mixed with bicarbonate of soda. When you pour a little cabbage water into each glass, the first liquid remains violet, the second turns red and the third green. The violet cabbage dye has the property of turning red in acid liquids and green in alkaline. In neutral water it does not change colour. In chemistry one can find out whether a liquid is acid or alkaline by using similar detecting liquids (indicators).

16

Violet becomes red

If you ever come across an anthill in the woods, you can there and then do a small chemical experiment. Hold a violet flower, e.g. a bluebell, firmly over the ants. The insects feel threatened and spray a sharp-smelling liquid over the flower. The places hit turn red.

The ants make a corrosive protective liquid in their hind quarters. You notice it if an ant nips you, though it is generally quite harmless. Since the flower turns red where the drops fall, you know that they are acid. The acid is called formic acid.

17

Invisible ink

If you ever want to write a secret message on paper, simply use vinegar, lemon, or onion juice, as the invisible ink. Write with it as usual on white writing paper. After it dries the writing is invisible. The person who receives the letter must know that the paper has to be held over a candle flame: the writing turns brown and is clearly visible.

Vinegar, and lemon or onion juice, cause a chemical change in the paper to a substance similar to cellophane. Because its ignition temperature is lower than that of the paper, the parts written on singe.

Bleached rose

A piece of sulphur is ignited in a jam jar. Since a pungent vapour is produced, you should do the experiment out-of-doors. Hold a red rose in the jar. The colour of the flower becomes visibly paler until it is white.

When sulphur is burned, sulphur dioxide is formed. As well as its germicidal action in sterilization, the gas has a bleaching effect, and the dye of the flower is destroyed by it. Sulphur dioxide also destroys the chlorophyll of plants, which explains their poor growth in industrial areas, where the gas pollutes the air.

19

Transfer pictures

Photos and drawings from newspapers can be copied easily. Mix two spoonfuls of water, one spoonful of turpentine and one spoonful of liquid detergent and dab this liquid with a sponge on the newspaper page. Lay a piece of writing paper on top, and after vigorous rubbing with a spoon the picture is clearly transferred to the paper.

Turpentine and liquid detergent when mixed form an emulsion which penetrates between the dye and oil particles of the dry printing ink and make it liquid again. Only newspaper printing ink can be dissolved, though. The glossy pictures in magazines contain too much lacquer, which is only dissolved with difficulty.

20

Sugar fire

Place a piece of cube sugar on a tin lid and try to set it alight. You will not succeed. However, if you dab a corner of the cube with a trace of cigarette ash and hold a burning match there, the sugar begins to burn with a blue flame until it is completely gone.

Cigarette ash and sugar cannot be separately ignited, but the ash initiates the combustion of the sugar. We call a substance which brings about a chemical reaction, without itself being changed, a catalyst.

21

Jet of flame

Light a candle, let it burn for a while, and blow it out again. White smoke rises from the wick. If you hold a burning match in the smoke, a jet of flame shoots down to the wick, and it relights.

After the flame is blown out the stearin is still so hot that it continues to evaporate and produce a vapour. But as this is combustible, it can be relighted at once by a naked flame. The experiment shows that solid substances first become gaseous at the surface before they will burn in a supply of oxygen.

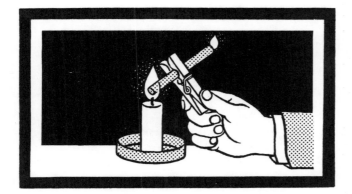

Roll a thin piece of tin foil round a pencil to make a tube about four inches long, and hold it with one end in the middle of a candle flame. If you hold a burning match at the other end of the tube, a second flame will be lit there.

Like all solid and liquid fuels, stearin produces combustible gases when heated, and these accumulate inside a flame. They burn, with the oxygen of the air, in the outer layer and tip of the flame. The unburnt stearin vapour in the middle can be drawn off, like town gas from the gas works.

Fix two plastic bags to the ends of a piece of wooden beading about 18 inches long and let it swing like a balance on a drawing pin. Pour some bicarbonate of soda and some vinegar into a glass. It begins to froth, because a gas is escaping. If you tilt the glass over one of the bags, the balance falls.

The gas which is given off during the chemical reaction is carbon dioxide. It is heavier than air, so it can be poured into the bag and weighed. If you were to fill a balloon with the gas it would never rise, and for this purpose other gases are used, which are lighter than air.

24

Fire extinguisher

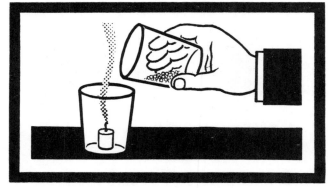

Light a candle stump in an empty glass, and mix in another glass — as in the previous experiment — a teaspoonful of bicarbonate of soda with some vinegar and let it froth. If you tilt the glass over the candle, the flame goes out.

The carbon dioxide formed in the chemical reaction in the top glass displaces the air needed for the flame, because it is heavier, and because it is noncombustible the flame is smothered. Many fire extinguishers work in the same way: the sprayed foam consists of bubbles filled with carbon dioxide. It surrounds the flame and blocks the supply of oxygen.

25

Burning without a flame

Press a handful of steel wool firmly into a glass tumbler and moisten it. Invert the the tumbler over a dish containing water. At first the air in the tumbler prevents the water entering, but soon the level of water in the dish becomes lower while that in the glass rises.

After the steel wool is moistened, it begins to rust. The iron combines with the oxygen in the air, and we call this process combustion or oxidation. Since the air consists of about one-fifth oxygen, the water rises in the tumbler until after some hours it fills one-fifth of the space. However, an imperceptible amount of heat is set free in the process.

Would you have thought that even iron could be made to burn with a flame? Twist some fine steel wool round a small piece of wood and hold it in a candle flame. The metal begins to blaze and scatter sparks like a sparkler.

The oxidation, which was slow in the previous experiment, is rapid in this case. The iron combines with the oxygen in the air to form iron oxide. The temperature thus produced is higher than the melting point of iron. Because of the falling red-hot particles of iron it is advisable to carry out the experiment in a basin.

Put a piece of aluminium foil with a copper coin on it into a glass of water, and let it stand for a day. After this the water looks cloudy and at the place where the coin was lying the aluminium foil is perforated.

This process of decomposition is known as corrosion. It often occurs at the point where two different metals are directly joined together. With metal mixtures (alloys) it is particularly common if the metals are not evenly distributed. In our experiment the water becomes cloudy due to dissolved aluminium. A fairly small electric current is also produced in this process.

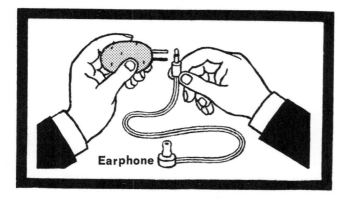

Earphone

Stick a finger-length piece of copper wire and a piece of non-copper wire into a raw potato. You can use a nail or straightened paper-clip for the non-copper wire. Hold the wires against the plug of a pocket radio earphone so that one wire is above the black insulation band of the plug and one is below. You will be able to hear a distinct crackling. The noise is caused by electric current. Similar to a flashlight battery, the potato and wires produce electricity although only a tiny amount. In a chemical reaction, electrons flow in a circuit from one metal to the other through the sap of the potato and the earphone. We call this circuit a galvanic cell.

Place several copper coins and standard zinc coated steel washers of the same size alternately above one another, and between each metal pair insert a piece of blotting paper soaked in salt water. Electrical energy, which you can detect, is set free. Wind thin, covered copper wire about 50 times round a compass, and hold one of the bare ends on the last coin and one on the last iron washer. The current causes a deflection of the compass needle.
In a similar experiment the Italian physicist Volta obtained a current. The salt solution acts on the metal like the sap in the potato in the previous experiment.

30

Graphite conductor

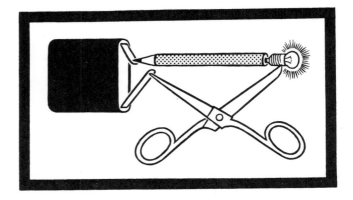

Connect a torch bulb with a battery by means of a pair of scissors and a pencil. The bulb lights up.

From the long tongue of the battery, the negative pole, the current flows through the metal of the scissors to the lamp. It makes it glow, and flows through the graphite shaft to the positive pole of the battery. Therefore graphite is a good conductor: so much electricity flows even through a pencil "lead" on paper, that you can hear crackling in earphones.

31

Mini-microphone

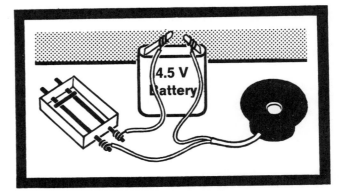

Push two pencil leads through the short sides of a matchbox, just above the base. Scrape off some of the surface, and do the same with a shorter lead, which you lay across the top. Connect the microphone with a battery and earphone in the next room. (You can take the earphone from a transistor radio.) Hold the box horizontal and speak into it. Your words can be heard clearly in the earphone.

The current flows through the graphite "leads". When you speak into the box, the base vibrates, causing pressure between the "leads" to alter and making the current flow unevenly. The current variations cause vibrations in the earphone.

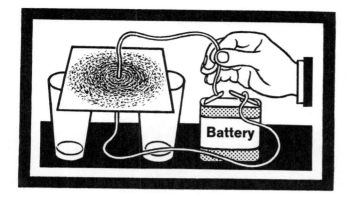

Push a length of copper wire through a piece of cardboard laid horizontally and connect the ends of the wire to a battery. Scatter iron filings on to the cardboard and tap it lightly with your finger. The iron filings form circles round the wire.

If a direct current is passed through a wire or another conductor, a magnetic field is produced round it. The experiment would not work with an alternating current, in which the direction of the current changes in rapid sequence, because the magnetic field would also be changing continuously.

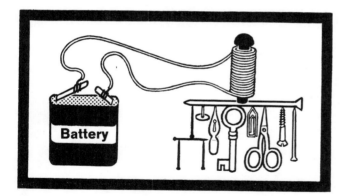

Wind one to two yards of thin insulated wire on to an iron bolt and connect the bare ends of the wire to a battery. The bolt will attract all sorts of metal objects.

The current produces a field of force in the coil. The tiny magnet particles in the iron become arranged in an orderly manner, so that the iron has a magnetic north and south pole. If the bolt is made of soft iron, it loses its magnetism when the current is switched off, but if it is made of steel it retains it.

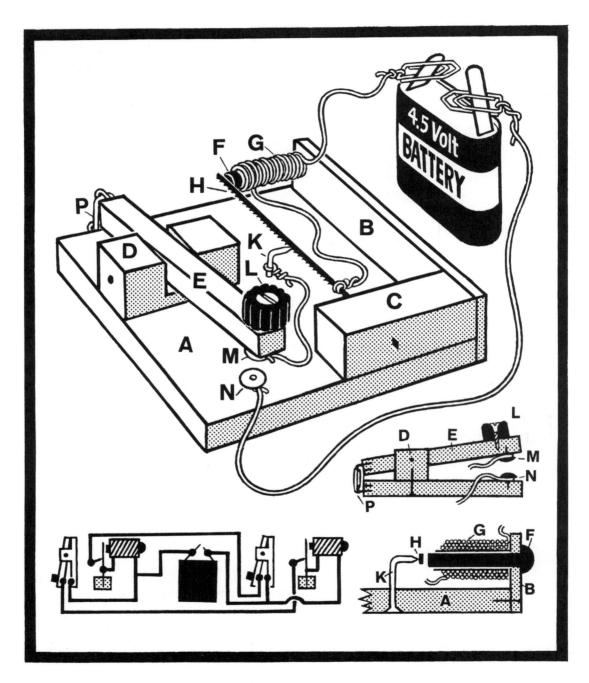

Electro-buzzer

Nail board B and wooden blocks C and D onto board A (about 5 x 5 inches). Push an iron bolt F through a hole bored in B. Wind covered copper wire G 100 times round the bolt and connect the ends to a battery and to H respectively. Bore a hole through block C and wedge the fretsaw blade H firmly into it so that its end is a short distance from bolt F. Hammer a long nail K through A and bend it so that its point rests in the middle of the saw blade. Oil the point of the nail. Use a piece of beading E as a key, with a rubber band P as spring and drawing pins M and N as contacts. Join all the parts with connecting wire (remove the insulation).

If you press the key down, you connect the electric circuit, bolt F becomes magnetic and attracts H. At this moment the circuit is broken at K and the bolt loses its magnetism. H jumps back and reconnects the current. This process is repeated so quickly that the saw blade vibrates and produces a loud buzz. If you wish to do morse signalling with two pieces of apparatus, you must use three leads as in the lower circuit diagram.

Light fan

Hold a light-coloured rod between your thumb and forefinger and move it quickly up and down in neon light. You do not see, as you might expect, a blurred, bright surface, but a fan with light and dark ribs.

Neon tubes contain a gas, which flashes on and off 50 times a second because of short breaks in alternating current. The moving rod is thrown alternatively into light and darkness in rapid sequence, so that it seems to move by jerks in a semi-circle. Normally the eye is too slow to notice these breaks in illumination clearly. In an electric light bulb the metal filament goes on glowing during the short breaks in current.

Static Electricity
Clinging balloons

Blow up some balloons, tie them up and rub them for a short time on a woollen pullover. If you put them on the ceiling, they will remain there for hours.

The balloons become electrically charged when they are rubbed, that is, they remove minute, negatively charged particles, called electrons, from the pullover. Because electrically charged bodies attract those which are uncharged, the balloons cling to the ceiling until the charges gradually become equal. This generally takes hours in a dry atmosphere because the electrons only flow slowly into the ceiling, which is a poor conductor.

Pepper and salt

Scatter some coarse salt onto the table and mix it with some ground pepper. How are you going to separate them again? Rub a plastic spoon with a woollen cloth and hold it over the mixture. The pepper jumps up to the spoon and remains sticking to it.

The plastic spoon becomes electrically charged when it is rubbed and attracts the mixture. If you do not hold the spoon too low, the pepper rises first because it is lighter than the salt. To catch the salt grains, you must hold the spoon lower.

38

Coiled adder

Cut a spiral-shaped coil from a piece of tissue paper about 4 inches square, lay it on a tin lid and bend its head up. Rub a fountain pen vigorously with a woollen cloth and hold it over the coil. It rises like a living snake and reaches upwards.

In this case the fountain pen has taken electrons from the woollen cloth and attracts the uncharged paper. On contact, the paper takes part of the electricity, but gives it up immediately to the lid, which is a good conductor. Since the paper is now uncharged again, it is again attracted, until the fountain pen has lost its charge.

39

Water bow

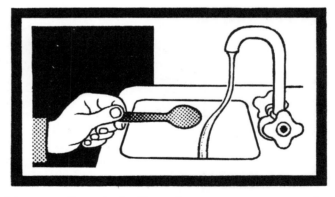

Once more rub a plastic spoon with a woollen cloth. Turn the water tap on gently and hold the spoon near the fine jet. At this point, the jet will be pulled towards the spoon in a bow.

The electric charge attracts the uncharged water particles. However, if the water touches the spoon, the spell is broken. Water conducts electricity and draws the charge from the spoon. Tiny water particles suspended in the air also take up electricity. Therefore experiments with static electricity always work best on clear days and in centrally heated rooms.

Hostile balloons

Blow two balloons right up and join them with string. Rub both on a woollen pullover and let them hang downwards from the string. They are not attracted, as you might expect, but float away from each other.

Both balloons have become negatively charged on rubbing because they have taken electrons from the pullover, which has now gained a positive charge. Negative and positive charges attract each other, so the balloons will stick to the pullover. Similar charges, however, repel one another, so the balloons try hard to get away from each other.

Shooting puffed rice

Charge a plastic spoon with a woollen cloth and hold it over a dish containing puffed rice. The grains jump up and remain hanging on the spoon until suddenly they shoot wildly in all directions.

The puffed rice grains are attracted to the electrically charged spoon and cling to it for a time. Some of the electrons pass from the spoon into the puffed rice, until the grains and the spoon have the same charge. Since, however, like charges repel one another, we have this unusual drama.

42

Simple electroscope

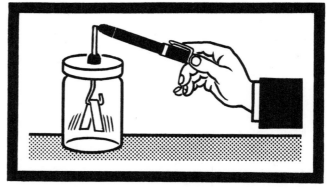

Bore a hole through the lid of a jam jar and push a piece of copper wire bent into a hook through it. Hang a folded strip of aluminium foil over the back. If you hold a fountain pen, comb, or similar object which has been electrically charged by rubbing on the top of the wire, the ends of the strip spring apart.

On contact with a charged object, electrical charges flow through the wire to the ends of the strip. Both now have the same charge and repel one another according to the strength of the charge.

43

Electrical ball game

Fix a piece of aluminium foil cut into the shape of a footballer on the edge of a phonograph record, rub the record vigorously with a woollen cloth and place it on a dry glass. Put a tin can about two inches in front of the figure. If you hold a small aluminium-foil ball on a thread between them, it swings repeatedly from the figure to the can and back.

The electric charge on the record flows into the aluminium-foil figure and attracts the ball. It becomes charged, but is immediately repelled because the charges become equal, and goes to the can, where it loses its electricity. This process is repeated for a time.

Electric fleas

Rub a long-playing record with a woollen cloth and place it on a glass. If you toss some small aluminium foil-balls on to the record, they will jump away from one another in a zig-zag motion. If you then move the balls together with your fingers, they will hop fiercely away again.

The electricity produced on the record by rubbing is distributed in irregular fields. The balls take up the charge and are repelled, but are again attracted to fields with the opposite charge. They will also be repelled when they meet balls with the same charge.

Puppet dance

Lay a pane of glass across two books, with a metal plate underneath. Cut out dolls an inch or so high from tissue paper. If you rub the glass with a woollen cloth, the dolls underneath begin a lively dance. They stand up, turn round in a circle, fall, and spring up again.

The glass becomes electrically charged when it is rubbed with the wool, attracts the dolls, and also charges them. Since the two like charges repel each other, the dolls fall back on the plate, give up their charge to the metal and are again attracted to the glass.

46

High voltage

Place a flat baking tray on a dry glass, rub a blown-up balloon vigorously on a woollen pullover and place it on the tray. If you put your finger near the edge of the tray, a spark jumps across.

A voltage equalization occurs between the metal and the finger. Although the spark is discharged with several thousand volts, it is just as harmless as the sparks produced when you comb your hair. An American scientist discovered that a cat's fur must be stroked 9,200,000,000 times to produce a current sufficient to light a 75-watt bulb for a minute.

47

Flash of lightning

Place a metal slice on a dry glass, and on it a piece of hard foam plastic which you have rubbed well on your pullover. If you hold your finger near the handle of the slice, a spark jumps across.

When the negatively charged plastic is placed on the slice, the negative electric particles in the metal are repelled to the end of the handle, and the voltage between it and the finger becomes equalized. Plastic materials can become strongly charged. In warehouses, for example, metal stands for rolls of plastic are earthed because otherwise they often spark when they are touched by the personnel.

Electric light

In many homes there is a voltage tester, generally in the form of a screwdriver. In its handle there is, amongst other things, a small neon tube which you can easily remove. Hold one metal end firmly and rub the other on a piece of hard foam plastic which may be used for insulation. The lamp begins to glow as it is rubbed to and fro, and you can see this particularly clearly in the dark.

Since the plastic is soft, its layers are rubbed against one another by the movement of the lamp and become strongly charged with electricity. The electrons collect on the surface, flow through the core of the tiny lamp, which begins to glow and into the body.

The ancient Greeks had already discovered that amber attracted other substances when it was rubbed. They called the petrified resin 'electron'. The power which has caused such fundamental changes in the world since then therefore gets its name — electricity.

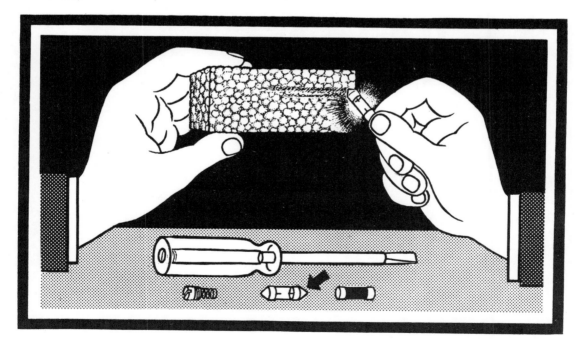

Lay a sheet of drawing paper over a magnet — of course you already know how to make a magnet — and scatter iron filings on it. Tap the paper lightly, and a pattern forms.

The filings form into curved lines and show the direction of the magnetic force. You can make the pattern permanent. Dip the paper into melted candle wax and let it cool. Scatter the iron filings on it. If you hold a hot iron over the paper after the formation of the magnetic lines, the field lines, the pattern will be fixed.

North

Hold a soft iron bar pointing to the north and sloping downwards, and hammer it several times. It will become slightly magnetic.

The earth is surrounded by magnetic field lines, which meet the earth in Great Britain and North America at an angle between 60° and 80°. When the iron is hammered, its magnet particles are affected by the earth's magnetic field lines and point to the north. In a similar way, tools sometimes become magnetic for no apparent reason. If you hold a magnetized bar in an east-west direction and hammer it, it loses its magnetism.

51

Magnetic or not ?

Many iron and steel objects are magnetized without one realizing it. You can detect this magnetism with a compass.

If a rod is magnetized, it must, like the compass needle, have a north and south pole. Since two unlike poles attract and two like poles repel, one pole of the needle will be attracted to the end of the bar and the other repelled. If the bar is not magnetized, both poles of the needle are attracted to the end.

52

Compass needle

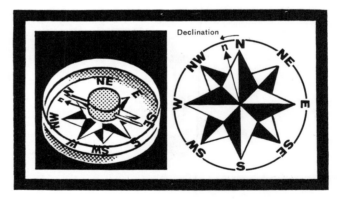

Stroke a sewing needle with a magnet until it is magnetized and push it through a cork disk. Put the needle into a transparent plastic lid containing water and it turns in a north-south direction. Stick a paper compass card under the lid.

The needle points towards the magnetic North pole of the earth. This lies in North Canada, and is not to be confused with the geographical North Pole, round which our earth rotates. The deviation (declination) of the magnetic needle from the true north is 8° in London and 15° in New York (in a westerly direction) and 1° in Chicago and 15° in Los Angeles, (in an easterly direction).

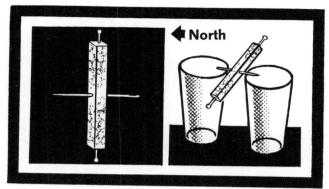

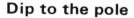

Magnetize two steel pins so that their points attract each other strongly. Push them into the ends of a piece of foam plastic about as thick as a pencil and balance this by means of a sewing needle over two tumblers (by shifting the pins and pulling off pieces of plastic). If you allow this compass to swing in a north-south direction, it will come to rest with the end facing north sloping downwards.

The compass needle comes to rest parallel to the magnetic field lines which span the earth from pole to pole. This deviation (dip) from the horizontal is $67°$ in London, $72°$ in New York, $60°$ in Los Angeles and at the magnetic poles of the earth $90°$.

Make two ducks from paper doubled over and glued and push a magnetized pin into each one. Place the ducks on cork disks in a dish of water. After moving around they line up with their beaks or tail tips together in a north-south direction.

The ducks approach each other along the magnetic field lines. Their movement is caused by different forces: the attraction of unlike magnetic poles, the repelling effect of like poles, and the earth's magnetism. Set the magnets so that two poles which will be attracted are placed in the beaks.

You can immerse a pocket handkerchief in water, without it getting wet: stuff the handkerchief firmly into a tumbler and immerse it upside down in the water.

Air is certainly invisible, but it nevertheless consists of minute particles which fill the available space. So air is also enclosed in the upturned glass, and it stops the water entering. If, however, you push the glass deeper, you will see that some water does enter, due to the increasing water pressure, which compresses the air slightly. Diving bells and caissons, used under water, work on the same principle.

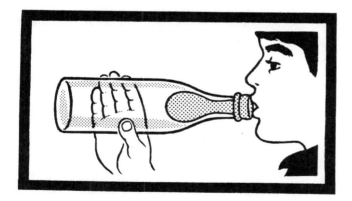

Do you believe that it is always possible to blow an ordinary balloon right up? You will be surprised: push a balloon into a bottle and stretch its mouth-piece over the opening. Blow hard into the balloon. It is only possible to stretch the rubber before your breath runs out.

As the pressure of the air in the balloon increases, so does the counter-pressure of the air enclosed in the bottle. It is soon so great that the breathing muscles in your thorax are not strong enough to overcome it.

57

Air lock

Place a funnel with not too wide a spout into the mouth of a bottle and seal it with plasticine so that it is airtight. If you pour some water into the funnel, it will not flow into the bottle.

The air enclosed in the bottle prevents the water entering. On the other hand, the water particles at the mouth of the funnel, compressed like a skin by surface tension, do not allow any air to escape. Close one end of a straw, push the other end through the funnel, lift your finger, and the water flows at once into the bottle. The air can now escape through the straw.

58

Hanging water

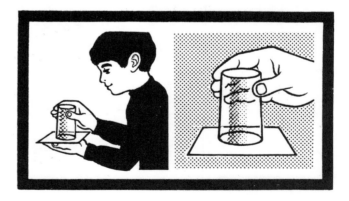

Fill a glass to overflowing with water and lay a postcard on it. Support the card with one hand, turn the glass upside down and remove your hand from the card. It remains on the glass, and allows no water to escape.

With a glass of normal height, a weight of water of about 2 ounces presses on each square inch of card. On the other hand the pressure of air from below is about one-hundred times as great on each square inch, and presses the card so firmly against the glass that no air can enter at the side and so no water can flow out.

Lay a cigar-box lid over the edge of a smooth table. Spread an undamaged sheet of newspaper and smooth it firmly on to the lid. Hit the projecting part of the lid hard with your fist. It breaks, without the paper flying up.

The lid is only slightly tilted when it is hit. In the space formed between the lid, newspaper and table, the air cannot flow in quickly enough, so that there is a partial vacuum, and the normal air pressure above holds the lid still as if it were in a screw clamp.

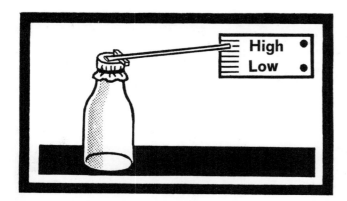

Stretch a piece of balloon rubber over the mouth of a milk bottle, stick a straw on top of it and put a matchstick between straw and rim of the bottle. As the air pressure varies daily according to the state of the weather, the end of the straw moves up and down.

When the air pressure is higher in fine weather, the rubber is pressed inwards, and the end of the pointer rises. When the air pressure falls, the pressure on the rubber is reduced, and the pointer falls. Because the air in the bottle will expand if it is heated, the barometer should be placed in a spot where the temperature will remain constant.

Weather frog

A tree frog made of paper will climb up and down a ladder like a real weather frog and predict the weather. Bend a 2½-inch-long strip of metal into a U-shape and bore through it so that a sewing needle can be turned easily when inserted through the holes. The needle is made able to grip by heating, and the frog, made from green paper, is fixed on to it by a thin wire. Stick the metal strip firmly on to the middle of the wall of a four-inch-high jar, and at the side a cardboard ladder. Wind a thread round the needle, with a small counterweight at the end. Stick a paper disk on a piece of plastic foil, and draw the other end of the thread through the middle. The foil is stretched over the mouth of the jar so as to be smooth and air-tight, the thread is tightly knotted, and the hole sealed.

When the air pressure is high (fine weather) the plastic foil is pressed inwards and the frog climbs up. When the pressure is low (bad weather) the pressure on the foil is less and the frog climbs back down.

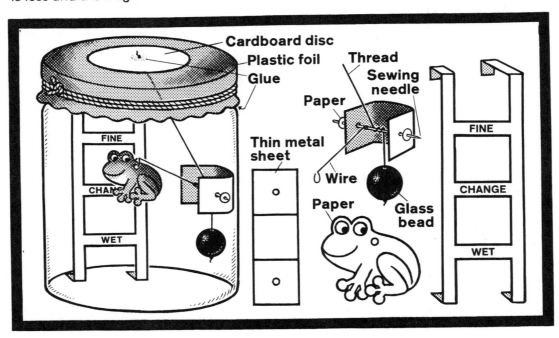

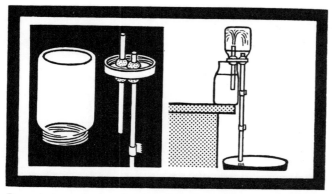

Punch two holes in the lid of a jam jar and push a plastic straw a distance of two inches through one. Fix three more straws together with adhesive tape and push through the other hole. Seal the joints with warm plasticine. Screw the lid to the jar, which should contain some water, turn it upside down and let the short straw dip into a bottle full of water: a fountain of water rises into the upper jar until the bottle is empty.

The water pours out through the long tube, and the air pressure in the jar becomes less. The air outside tries to get in and pushes the water from the bottle.

It is simple, using air, to lift matches from the table into their box. Hold the case between your lips and lower it over the matches. Draw a deep breath, and the matches hang on to the bottom of the case as though they were stuck on.

By drawing in breath you produce a dilution of the air, in the case. Air pressure pushes the matches from underneath towards the opening. Even a single match can be raised in this way, if the air is drawn in sharply.

64

Shooting backwards

Hold an empty bottle horizontal and place a small paper ball just inside its neck. Try to blow the ball into the bottle. You cannot! Instead of going into the bottle, the ball flies towards your face.

When you blow, the air pressure in the bottle is increased, and at the same time there is a partial vacuum just inside the neck. The pressures become equalized so that the ball is driven out as from an airgun.

65

Blowing trick

Place a playing card on a wine glass so that at the side only a small gap remains. Lay a large coin (half a dollar or 10 new pence) on the card. The task is to get the coin into the glass. Anybody who does not know the trick will try to blow the coin into the gap from the side without success. The experiment only works if you blow once quickly into the mouth of the glass. The air is trapped inside and compressed. The increased pressure lifts the card and the coin slides over it and into the glass.

Compressed air rocket

Bore a hole through the cap of a plastic bottle, push a plastic drinking straw through it and seal the joints with adhesive. This is the launching pad. Make the rocket from a four-inch-long straw, which must slide smoothly over the plastic straw. Stick coloured paper triangles for the tail unit at one end of the straw, and at the other end plasticine as the head. Now push the plastic tube into the rocket until its tip sticks lightly into the plasticine. If you press hard on the bottle the projectile will fly a distance of 10 yards or more.

When you press the plastic bottle, the air inside is compressed. When the pressure is great enough, the plastic straw is released from the plug of plasticine, the released air expands again, and shoots off the projectile. The plasticine has the same function as the discharge mechanism in an airgun.

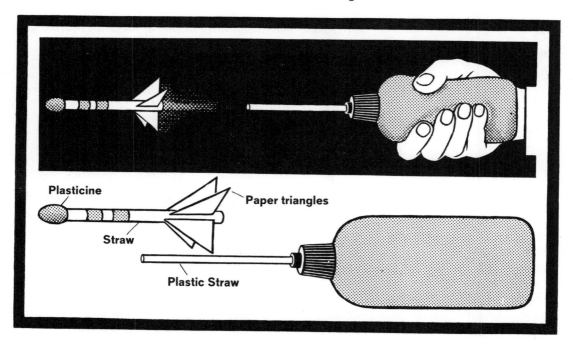

Plasticine

Paper triangles

Straw

Plastic Straw

67

Egg blowing

Place two porcelain egg-cups one in front of the other, with an egg in the front one. Blow hard from above on to the edge of the filled cup. Suddenly the egg rises, turns upside down and falls into the empty cup. Because the egg shell is rough, it does not lie flat against the smooth wall of the egg-cup. Air is blown through the gap into the space under the egg, where it becomes compressed. When the pressure of the air cushion is great enough, it lifts the egg upwards.

68

Curious air currents

If you stand behind a tree trunk or a round pillar on a windy day, you will notice that it offers no protection, and a lighted match will be extinguished. A small experiment at home will confirm this: blow hard against a bottle which has a burning candle standing behind it, and the flame goes out at once.
The air current divides on hitting the bottle, clings to the sides, and joins up again behind the bottle with its strength hardly reduced. It forms an eddy which hits the flame. You can put out a lighted candle placed behind two bottles in this way, if you have a good blow.

Lay a postcard bent lengthways on the table. You would certainly think that it would be easy to overturn the card if you blew hard underneath it. Try it! However hard you blow, the card will not rise from the table. On the contrary, it clings more firmly.

Daniel Bernoulli, a Swiss scientist of the eighteenth century, discovered that the pressure of a gas is lower at higher speed. The air stream produces a lower pressure under the card, so that the normal air pressure above presses the card on to the table.

Push three pins into the middle of a piece of wood and lay a coin (5 new pence or 25 cents) on top of them. You can make a bet! Nobody who does not know the experiment will be able to blow the coin off the tripod.

The metal cannot hold the gust of air on its narrow, smooth edges. The gust shoots through under the coin and reduces the air pressure, forcing the coin more firmly on to the pins. But if you lay your chin on the wood just in front of the coin and blow with your lower lip pushed forward, the air hits the underside of the coin directly and lifts it off.

71

Trapped ball

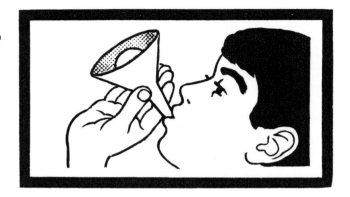

Place a table tennis ball in a funnel, hold it with the mouth sloping upwards, and blow as hard as you can through the spout. You would hardly believe it, but nobody can manage to blow the ball out.

The air current does not hit the ball, as one would assume, with its full force. It separates and pushes through the places where the ball rests on the funnel. At these points the air pressure is lowered according to Bernoulli's law, and the external air pressure pushes the ball firmly into the mouth of the funnel.

72

Flying coin

Lay a sixpence or a dime four inches from the edge of the table and place a shallow dish eight inches beyond it. How can you blow the coin into the dish?

You will never do it if you blow at the coin from the front — on the false assumption that the air will be blown under the coin because of the unevenness of the table and lift it up. It will only be transferred to the dish if you blow once sharply about two inches horizontally above it. The air pressure above the coin is reduced, the surrounding air, which is at normal pressure, flows in from all directions and lifts the coin. It goes into the air current and spins into the dish.

Floating card

Many physical experiments seem like magic, but there are logical explanations and laws for all the strange occurrences. Stick a thumb tack through the middle of a halved postcard. Hold it under a cotton spool so that the pin projects into the hole and blow hard down the hole. If you manage to loosen the card, you really expect it to fall. In fact, it remains hovering under the spool.

This surprising result is explained by Bernoulli's law. The air current goes through at high speed between the card and the spool, producing a lower pressure, and the normal air pressure pushes the card from below against the spool. The ascent of an aeroplane takes place in a similar manner. The air flows over the arched upper surface of the wings faster than over the flat under-surface, and therefore the air pressure above the wings is reduced.

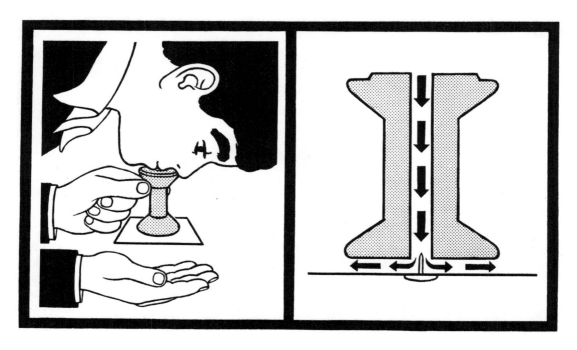

74

Wind funnel

Light a candle and blow at it hard through a funnel held with its mouth a little way from the flame. You cannot blow out the flame; on the contrary it moves towards the funnel.

When you blow through the funnel the air pressure inside is reduced, and so the air outside enters the space through the mouth. The blow air sweeps along the funnel walls: if you hold the funnel with the edge directly in front of the flame, it goes out. If you blow the candle through the mouth of the funnel, the air is compressed in the narrow spout, and extinguishes the flame immediately on exit.

75

Explosion in a bottle

Throw a burning piece of paper into an empty milk bottle and stretch a piece of balloon rubber firmly over the mouth. After a few moments, the rubber is sucked into the neck of the bottle and the flame goes out.

During combustion, part of the expanded, hot air escapes. After the flame goes out the diluted gas in the bottle cools and is compressed by the external pressure. The rubber is therefore stretched so much that the final pressure equalization only occurs if you break the bubble, causing a loud pop.

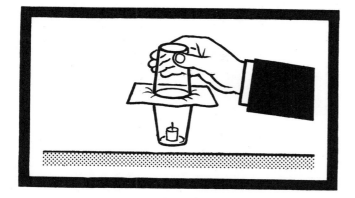

Light a candle stump in an empty tumbler, lay a sheet of damp blotting paper over the top and invert a second tumbler of the same size over it. After several seconds the flame goes out and the tumblers stick together.

During combustion the oxygen in both tumblers is used up – the blotting paper is permeable to air. Therefore the pressure inside is reduced and the air pressure outside pushes the tumblers together.

Place a coin in a dish of water. How can you get it out, without putting your hand in the water or pouring the water from the dish? Put a burning piece of paper in a tumbler and invert it on the dish next to the coin. The water rises into the tumbler and releases the coin.

During combustion the carbon contained in the paper, together with other substances, combines with the oxygen in the air to form carbon dioxide. The gas pressure in the tumbler is reduced by the expansion of the gases on heating and contraction on cooling. The air flowing in from outside pushes the water into the tumbler.

Heat
Bottle ghost

An empty wine bottle which has been stored in a cool place, has a ghost in it!
Moisten the rim of the mouth with water and cover it with a coin. Place your
hands on the bottle. Suddenly the coin will move as if by a ghostly hand.
The cold air in the bottle is warmed by your hands and expands, but is prevented
from escaping by the water between the bottle rim and the coin. However, when the
pressure is great enough, the coin behaves like a valve, lifting up and allowing the
warm air to escape.

Expanded air

Pull a balloon over the mouth of a bottle and place in a saucepan of cold water. If
you heat the water on a stove, the balloon is seen to fill with air.
The air particles in the bottle whirl around in all directions, thus moving further
apart, and the air expands. This causes an increased pressure, which escapes into
the balloon and causes it to distend. If you take the bottle out of the saucepan,
the air gradually cools down again and the balloon collapses.

80

Buddel thermometer

Pour some coloured water into a bottle. Push a drinking straw through a hole bored in the cork so that it dips into the water. Seal the cork with glue. If you place your hands firmly on the bottle, the water rises up the straw.

The air enclosed in the bottle expands on heating and presses on the water surface. The displaced water escapes into the straw and shows the degree of heating by its position. You can fix a scale on the side of the bottle.

81

Hot-air balloon

Roll a paper napkin (not too soft material) into a tube and twist up the top. Stand it upright and light the tip. While the lower part is still burning, the ash formed rises into the air. Take care! The air enclosed by the paper is heated by the flame and expands. The light, balloon-like ash residue experiences a surprising buoyancy because the hot air can escape, and the air remaining in the balloon becomes correspondingly lighter. Very fine napkins are not suitable for the experiment because the ash formed is not firm enough.

Take an empty, corked wine bottle, push as long an aluminium knitting needle as you can find into the bottle cork and let the other end project under slight pressure over the mouth of a second, uncorked bottle. Glue a paper arrow on to a sewing needle, making sure that it is balanced, and fix it between the knitting needle and the neck of the bottle. Place a candle so that the tip of the flame touches the middle of the needle and watch the arrow.

The arrow turns quite quickly some way to the right because the knitting needle expands on heating like other substances. With an ordinary steel knitting needle the arrow would only turn a little, because steel only expands half as much as aluminium. Since the aluminium is longer as well, the difference is still greater. The expansion is clearly visible in electricity power cables, which sag more in summer than in winter. If you take the candle away from the knitting needle, the arrow moves back.

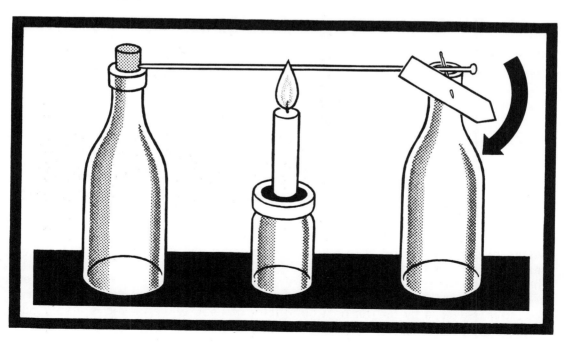

83

Exploding stone

You can explode large stones in the winter quite easily. Look for a flint that is well frozen through and pour boiling water over it. It breaks apart with cracks and bangs. The explosive effect is caused by the outer layers heating and expanding faster than the centre. The resulting tension causes the stone to burst. In the same way thick-walled glasses may explode if you pour hot liquids into them. Glass conducts heat poorly, so that the layers of glass expand by different amounts.

84

A clear case

Put spoons made of steel, silver, and plastic and a glass rod into a glass. Fix a dry pea at the same height on each handle with a dab of butter. In which order will the peas fall if you pour boiling water into the glass?

The butter on the silver spoon melts very quickly and releases its pea first. The peas from the steel spoon and the glass rod fall later, while that on the plastic spoon does not move. Silver is by far the best conductor of heat, while plastic is a very poor conductor, which is why saucepans, for example, often have plastic handles.

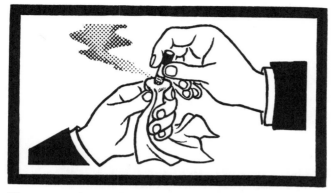

Place a coin under a cotton handkerchief and ask someone to press a burning cigarette on the cloth stretched over the coin. You need not be afraid of scorching the material, because only a harmless speck of ash will be left.

The experiment shows that the metal of the coin is a much better conductor of heat than the cotton fabric. On rapid pressure the heat of the burning cigarette is immediately conducted away by the coin. There is only enough heat to cause a small rise in temperature in the coin, and the cotton does not reach a high enough temperature to burn.

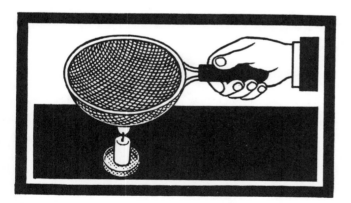

Hold a metal kitchen sieve in a candle flame. To your surprise the flame only reaches the wire net, but does not go through it.

The metal in the sieve conducts so much heat away that the candle wax vapour cannot ignite above the wire net. The flame only passes through the metal lattice if it is made to glow by strong heating. The miner's safety lamp works in the same way. A metal lattice surrounding the naked flame takes up so much heat that the gases in the mine cannot ignite.

Scenting coins

Three different coins lie in a plastic dish. You close your eyes while another person takes out one coin, holds it for several seconds in his closed hand, and puts it back. Now hold the coins one after the other briefly to your upper lip and find out immediately, to everyone's astonishment, which coin was taken from the dish. Since metals are very good conductors of heat, the coin warms up immediately in the hand. But plastic is a poor conductor, so hardly any heat is lost to the dish when the coin is put back. The upper lip is particularly sensitive and reveals the smallest temperature difference in the coins, so that you can detect the right one immediately. Before the trick is repeated it is a good idea to lay the coins on a cold stone floor to conduct away the heat.

Fire under water

Warm the base of a candle stump and stick it in a bowl. Fill the bowl with cold water up to the rim of the candle. If you light the wick it burns until it is under the surface of the water. Then the candle flame hollows out a deep funnel. An extremely thin wall of wax remains standing round the flame and stops the water from extinguishing it.

The water takes so much heat from the candle that its outer layer does not reach its melting point, and the wax there cannot evaporate and burn.

Paper saucepan

Do you believe that you can boil water in a paper cup over a naked flame or in the embers of a fire? Push a knitting needle through the rim of a paper cup containing some water, hang it between two upright bottles and light a candle under the cup. After a little while the water boils — but the cup is not even scorched.

The water removes the heat transferred to the paper and begins to boil at a temperature of 212°F or 100°C. The water does not get any hotter, so the paper does not reach the temperature which is necessary for it to burn.

Evaporation and vaporization
Jet boat

Leave the clean, scooped out skin of a pineapple on a radiator for approximately three hours, to dry. Then mould it into the shape of a ship. Fasten a candle stump underneath the boat and a blown out egg shell on top of the boat by sticking two wire paper clips through the skin. Close the hole at the front of the egg shell and fill it up with a little water. Put the boat into a tub full of water. If you light the candle, the water inside the egg should soon start to boil and a jet of steam will start to gush out of the shell. The steam will build up pressure and shoot out through the nozzle with a great force. This will propel the boat forward in the opposite direction to the steam.

Hovercraft

Place a tin lid on a hot-plate and heat it well (take care!). If you then let a few drops of water fall on the lid, you will observe a small natural phenomenon. The drops are suspended in the air like hovercraft and whizz hissing to and fro for a while.

On contact with the heated metal the water drops begin to evaporate at once on the underside. Since the steam escapes with great pressure, it lifts the drops into the air. So much heat is removed from the drops by the formation of steam that they do not even boil.

92

Rain in the room

Rain after sultry days makes the inside of the window pane suddenly sweat. You can distinguish the tiny water droplets through a magnifying glass. Where do they come from?

After it has been raining the air outside cools sharply because the water evaporates and thus uses heat. The warm air in the room, which is saturated with water vapour, especially from cooking, cools down only slowly on the window pane. But cold air cannot hold so much moisture as warm air, and therefore loses some of it on to the pane. It forms water droplets — exactly as when it is raining out-of-doors and moist, warm air meets cold air.

93

Weather station

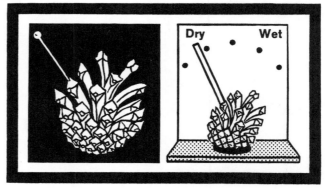

Fix a dry pine cone on to a small piece of wood with sealing wax or glue. Stick a pin into one of the central scales and place a straw over it. Put the cone out-of-doors, protected from the rain. The straw moves according to the state of the weather. Fix up a scale.

This simple hygrometer was built by nature. The pine cone closes when it is going to rain, to protect the seeds from damp. The outside of the scales absorbs the moisture in the air, swells up and bends — a process which you can also observe with a piece of paper which is wet on one side.

Hygrometer

Coat a strip of writing paper two inches long with glue and roll it onto a sewing needle. Stick a strip of shiny photographic paper about $\frac{1}{2}$ inch wide and one foot long onto its end so that its shiny surface faces the glue-covered side of the writing paper. The film strip is rolled round the needle like a clock spring. Punch a small hole through the middle of the bottom and lid of a furniture polish tin, and also air holes in the bottom. File off the metal projections formed. Push the needle through the central holes and stick the end of the film strip firmly to the side of the tin. Fix a paper pointer in front of the needle with a cork disk, and a bead behind it.

The gelatin layer of the photographic film expands – in contrast to the paper layer – with increased air humidity, causing it to wind up sharply, and move the pointer to the right. When the humidity of the air falls, the pointer returns to the left.

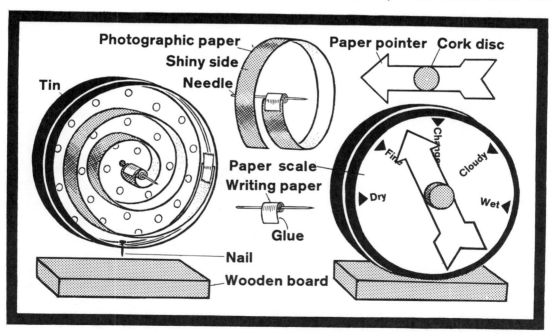

95

Water from the desert

We still read in the newspapers of people dying of thirst in the desert, but many of them could help themselves in this emergency. An experiment on a small scale in a sand box will show you how to do it. Dig a fairly deep hole and place a beaker in the middle. Spread a suitably sized piece of transparent plastic foil over the edge of the hole and lay a small stone in its centre so that it dips down to the beaker in the shape of a funnel. The edges are fixed firmly into the sand. Soon, especially in sunshine, small drops of water form on the underside of the foil. They become larger and larger and finally flow into the beaker.

The effect of the sun is to heat the ground strongly under the foil. The moisture held in the sand evaporates until the enclosed air is so saturated that small drops of water are deposited on the cooler foil. Even desert sand contains some moisture. If you also place cut up cactus plants into the hole, you will obtain enough water to survive.

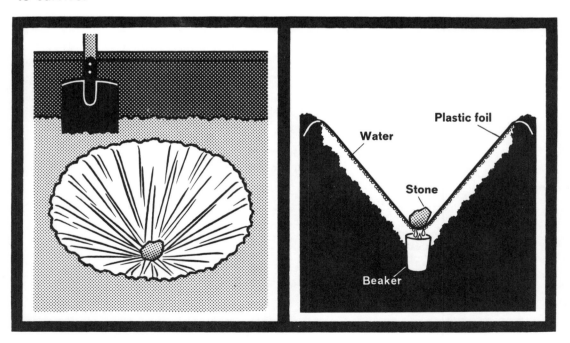

Bath game with a coin

Stretch a strip of cellophane (not plastic foil), 1 inch wide, tightly over a soup plate and fasten the ends with adhesive tape. Lay on the middle of the strip an average-sized coin and pour water into the dish up to about $\frac{1}{2}$ inch under the coin. The coin sinks slowly and reaches the water after several minutes.

The water vaporizes, the cellophane absorbs the water particles from the air and expands until it reaches the water. But strangely enough it soon begins to tighten again, and the coin rises again slowly to its original position.

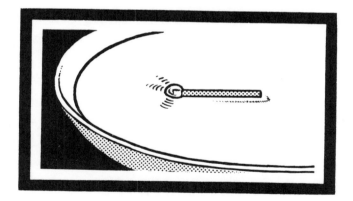

Steam boat

Break off the head of a match and drop some glue on to the end. If you place the match in a dish of water it moves jerkily forward.

The glue contains a solvent which evaporates to give a vapour. It puffs out from the drop in invisible little clouds, giving the match a small push each time. Eventually so much of the solvent has escaped that the glue becomes solid. In a dried drop of glue you can still see the residual solvent vapour as small bubbles.

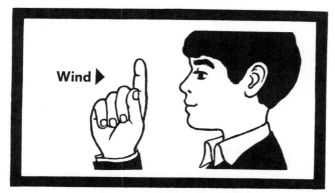

Cold and ice
Where is the wind coming from?

Moisten your finger and hold it straight up in the air. You will notice at once that one side of the finger is cold. This is the direction from which the wind is coming.

Heat is used up when a liquid vaporizes or evaporates. The wind accelerates the evaporation of the moisture on the finger and you will notice even with a weak air current the greater heat loss on the side facing the wind. Anybody who keeps on a wet bathing costume after a swim will shiver even in the heat. The water takes heat from the body as it evaporates.

Producing cold

With a rubber band fix a wad of cotton wool over the mercury bulb of a room thermometer. Note the temperature, damp the cotton wool with eau-de-cologne, and whirl the thermometer round on a string for a time. The temperature drops considerably. The alcohol in the eau-de-cologne evaporates quickly and so uses up heat. The draught caused by whirling the thermometer round accelerates the process and the heat consumption rises. In a refrigerator a chemical liquid evaporates in a container. The large amount of heat needed for this is taken from the food compartment.

100

Column of ice

Place an ink bottle filled to the brim with water in the freezing compartment of a refrigerator. Soon a column of ice will stick up out of the bottle.
Water behaves oddly: when warm water cools it contracts, but if the temperature falls below 39°F or 4°C, it suddenly begins to expand again. At 32°F or 0°C it begins to freeze, and in doing so increases its volume by one-eleventh. This is the reason why the ice sticks out of the bottle. If you had closed it, it would have cracked. Think about burst water pipes in winter and frost cracks on roads, in which water collected under the asphalt freezes.

101

Iceberg

Place a cube of ice in a tumbler and fill it to the brim with water. The ice cube floats and partly projects from the surface. Will the water overflow when the ice cube melts?
The water increases its volume by one-eleventh when it freezes. The ice is therefore lighter than water, floats on the water surface and projects above it. It loses its increased volume when it melts and exactly fills the space which the ice cube took up in the water. Icebergs, which are a danger to navigation, are therefore especially harmful because one only sees their tips above the water.

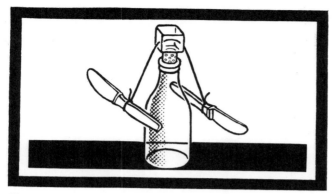

Cutting through ice

Place an ice cube on the cork of a bottle. Fix two objects of equal weight on a piece of wire, hang the wire over the ice and place the whole lot out of doors in frosty weather. After a certain time the wire will have cut through the ice without dividing it.

This trick of nature is explained by the fact that ice melts when it is subjected to pressure. Water is formed where the wire is resting, while it immediately freezes again above it. Skating is only made possible by slight melting of the ice under the moving surface, which reduces the friction.

Ice hook

Who can hook an icè cube from a bowl of water with a match? A trick makes it quite easy: place the match on the ice cube and scatter some salt over it. In no time the match is frozen solid, and you can lift it together with the ice cube from the dish.

Salt water does not freeze as easily as ordinary water, and scattering salt on ice makes it melt. The salt grains on the ice cube also do this. However, when a substance melts, heat is consumed at the same time. This heat is taken from the moisture under the match, where no salt fell, in this case — and it freezes.

Let a fine jet of water pour on a finger held about two inches under the tap. If you look carefully, you will see a strange wave-like pattern in the water.

If you bring your finger closer to the tap, the waves become continuously more ball-shaped, until the water jet resembles a string of pearls. It is so strongly obstructed by the finger that because of its surface tension – the force which holds the water particles together – it separates into round droplets. If you take your finger further away from the tap, the falling speed of the water becomes greater, and the drop formation is less clear.

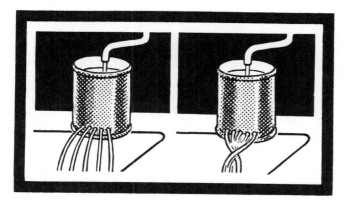

An empty two-lb. can is pierced five times just above the lower edge with a thin nail. The first hole should be just over an inch from the fifth. Place the tin under a running tap, and a jet will flow from each hole. If you move your finger over the holes, the jets will join together.

The water particles are attracted to one another and produce a force acting into the interior of the liquid, the surface tension. It is also this force which holds a water droplet together. In our experiment the force is particularly clear, and it diverts the jets into a sideways arc and knots them.

106

Mountain of water

Fill a dry glass just full with tap water, without any overflowing. Slide coins carefully into the glass, one after the other, and notice how the water curves above the glass.

It is surprising how many coins you can put in without the water spilling over. The water mountain is supported by surface tension, as though it is covered by a fine skin. Finally, you can even shake the contents of a salt cellar slowly into the glass. The salt dissolves without the water pouring out.

107

Ship on a High Sea

Place a half dollar or 10 new pence on the table, and on one side of the coin a small cork disk. How can you move the cork to the exact centre of the coin without touching it?

Pour water on to the coin — dropwise, so that it does not spill over — to form a water mountain over the surface. At first the force of gravity holds the cork on the edge of the slightly curved water surface. If you now pour on more water, the pressure of the water on the edge increases, while it remains constant on the top. So the cork moves up the hill to the middle, which is the region of lowest pressure.

Floating metal

Fill a bowl with tap water. Place small metal objects on blotting paper and carry it carefully into the dish with a fork. After a time the saturated blotting paper sinks, but the small objects remain floating.

Since metal is heavier than water, it should really fall to the bottom. The liquid particles are held together so strongly by a mysterious force, the surface tension, that they prevent the objects sinking. Surface tension is destroyed by soap.

Watertight sieve

Fill a milk bottle with water and fasten a piece of wire gauze about two inches square over its mouth with a rubber band. Place your hand over the top and turn the bottle upside down. If you take your hand away quickly, no water comes out. Where water comes into contact with air, it surrounds itself as though with a skin, because of its surface tension. Each opening in the wire gauze is so well sealed, that air can neither flow in nor water flow out. This also occurs with the fine holes of tenting material, which is made water-repellent by impregnation, and rain drops cannot get through because of their surface tension.

110

Rope trick

Knot a piece of string into a loop and allow it to float in a bowl of water. If you dip a match into the middle of the irregularly shaped loop, it immediately becomes circular.

The match has this magic power because it was previously dabbed with a little washing-up liquid. This spreads in all directions when the match is dipped into the water and penetrates between the water particles, which were held together like a skin by surface tension. This 'water skin' breaks in a flash from the place where the match is dipped in outwards. The liquid particles which are made to move push against the loop and make it rigid.

111

Speedboat

Split a match slightly at its lower end and smear some soft soap into the slit. If you place the match in a dish of tap water, it moves forwards quickly for quite a time. Several matches could have a race in a bath tub.

The soap destroys the surface tension of the water by degrees as it gradually dissolves. This causes a backward movement of the water particles, which produces as a reaction a forward movement of the match. With a drop of detergent instead of the soap the movement would be like a rocket.

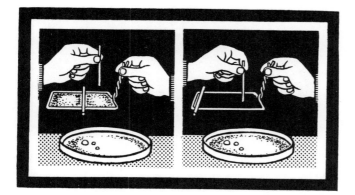

Make a rectangular frame about 1 x 3 inches out of thin wire. Place a straight piece of wire loosely over the centre. Dip the whole thing into washing-up liquid, so that a fine film stretches over it. If you pierce through one side, the piece of wire rolls backwards to the other end of the frame. The liquid particles attract one another so strongly that the soap film is almost as elastic as a blown-up balloon. If you break the cohesion of the particles on one side, the force of attraction on the other side predominates, the remaining liquid is drawn over and the wire rolls with it.

113

Soap bubbles

In each plastic detergent bottle that is thrown away there are still a thousand soap bubbles! Cut off the lower third of an empty detergent bottle and mix 10 teaspoonfuls of water with the detergent remaining in it. Bore a hole in the cap, push a straw through it, and a match into the nozzle. Put some of the liquid into the pipe and blow!
The liquid particles in the soap bubbles are compressed from outside and inside by surface tension. They hold together so strongly that they enclose the air flowing from the pipe and so take on the shape of the smallest surface, which is a sphere.

114

Bag of wind

Prevent a large soap bubble sitting on the pipe from flying off by closing the end of the straw with your finger. Hold it near a candle flame and then take your finger away. The flame leans to the side, while the soap bubble becomes smaller and vanishes.

Although a soap bubble film is generally less than one-thirty-thousandth of a inch thick, it is so strong that the air inside is compressed. When the end of the straw is released, the liquid particles contract to form drops again because of the surface tension, and thus push the air out.

115

Water rose

Cut out a flower shape from smooth writing paper, colour it with crayons and fold the petals firmly inwards. If you place the rose on water you will see the flower petals open in slow motion.

Paper consists mainly of plant fibres, which are composed of extremely fine tubes. The water rises in these so-called capillary tubes. The paper swells, and the petals of the synthetic rose rise up, like the leaves of a wilting plant when it is placed in water.

Fill a preserving jar, as tall as you can find, with water, stand a brandy glass in it and try to drop coins into the glass. It is very surprising that however carefully you aim, the coin nearly always slips away to the side.

It is very seldom possible to get the coin straight into the water. The very smallest slope is enough to cause a greater resistance of the water on the slanting under side of the coin. Because its centre of gravity lies exactly in the middle, it turns easily and drifts to the side.

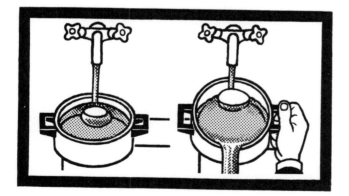

If you cool a boiled egg in the usual way under the water tap, you can make a surprising discovery. Hold the saucepan so that the water runs between the egg and the rim. If you now lean the saucepan to the other side the egg does not, as you would expect, roll down the bottom of the saucepan, but stays in the stream of water.

By Bernouilli's Law the pressure of a liquid or a gas becomes lower with increasing speed (see experiments 69 – 74). In the stream of water between the rim of the pan and the egg there is reduced pressure, and the egg is pressed by the surrounding water, which is at normal pressure, against the pan.

118
Buoyancy
Loss of weight

Tie a stone by means of a thread to a spring balance and note its weight. Does it in fact alter if you hang the stone in a jar of water?

If you lift up a large stone under the water when you are bathing you will be surprised at first by its apparently low weight. But if you lift it out of the water, you will see how heavy it actually is. In fact an object immersed in a liquid (or in a gas) loses weight. This is particularly obvious with a floating object. Look at the next experiment.

119
Archimedes' principle

Fill a container up to the brim with water and weigh it. Then place a block of wood on the water, and some of the liquid will spill out. Weigh again, to find out if the weight has altered.

The weight remains the same. The water spilt out of the container weighs exactly the same as the whole block of wood. The famous mathematician Archimedes discovered in about 250 B.C. that a body immersed in a liquid loses as much weight as the weight of liquid displaced by it. This apparent loss of weight is called buoyancy.

120

Water puzzle

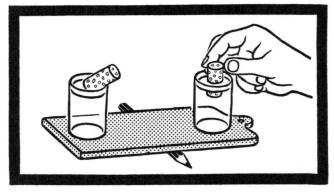

Lay a small tray or a wooden ruler over a six-sided pencil and place on it two jars filled with water balanced as on a pair of scales. What happens if you immerse a cork into the water in one jar, while placing a cork of the same size on the water in the other jar? Does one side of the balance become heavier, and if so, which side?

The balance leans to the side where you immerse the cork in the jar. That is, this side increases in weight by exactly as much as the weight of water displaced by the cork. The other jar only becomes as much heavier as the weight of the cork itself.

121

Mysterious water level

Place a half-dollar, ten new pence, or a penny in a match box and float it in a glass of water. Mark the level of the water on the side of the glass. Will it rise or fall if you take the coin from the box and lower it into the water? Just think about it first!

The water level falls. Since the coin is almost ten times heavier than water, the box containing the coin also displaces, because of its larger volume, ten times more water than the coin alone. This takes up, in spite of its greater weight, only a small volume and so displaces only a small amount of water.

Volcano under water

Fill a small bottle full of hot water and colour it with ink. Lower the bottle by means of a string into a preserving jar containing cold water. A coloured cloud, which spreads to the surface of the water, rises upwards out of the small bottle like a volcano.

Hot water occupies a greater volume than cold because the space between the water particles is increased on heating. It is, therefore, lighter and experiences buoyancy. After some time the warm and cold water mix and the ink is evenly distributed.

123

Suspending an egg

Half fill a jar with water and dissolve plenty of salt in it. Now add as much water again, pouring carefully over a spoon so that the two liquids do not mix. An egg placed in the jar remains suspended as though bewitched in the middle.

Since the egg is heavier than tap water, but lighter than salt water, it sinks only to the middle of the jar and floats on the salt water. You can use a raw potato instead of the egg. Cut a roundish 'magic fish' from it, and make fins and eyes from coloured cellophane.

124

Dance of the moth balls

Add some vinegar and bicarbonate of soda to some water in a jar. Toss several moth balls, which you can colour beforehand with a crayon to make the experiment more fun, into the bubbling bath. After a time the balls dance merrily up and down. Since the moth balls are a little heavier than water, they sink to the bottom of the jar. The carbon dioxide freed by the chemical reaction between vinegar and bicarbonate of soda collects in bubbles on the balls and lifts them slowly to the surface of the water. There the bubbles escape, the balls sink again, and the performance is repeated.

125

Pearl diver

Stick a match about one-tenth of an inch deep into a coloured plastic bead and shorten it so that its end floats exactly on the surface of the water when the bead is placed in a milk bottle full of water. Close this with a plastic cap. By changing the pressure on the cap, the bead can be made to rise and fall as though bewitched. Plastic is only a little heavier than water. The match and the air in the hole of the bead give it just enough buoyancy for it to float. The pressure of the finger is conducted through the water and compresses the air in the bead. Thus it no longer has sufficient buoyancy and sinks.

The yellow submarine

Cut a small boat out of fresh orange peel and make portholes on it with a ballpoint pen. Place the boat in a bottle filled with water and close it with a plastic cap. If you press on the cap, the boat rises and falls according to the strength of the pressure. Minute air bubbles in the porous fruit peel make it float. By the pressure of the finger, which is transmitted through the water, the bubbles are slightly compressed, so their buoyancy is less, and the boat goes to diving stations. Since the yellow of the peel is heavier than the white, the submarine floats horizontally.

You can accompany the submarine by several 'frogmen'. Simply toss broken-off match heads with it into the bottle. They float, because air is also contained in their porous structure. If the air bubbles are made smaller by the transmitted water pressure, the match heads dive deeper too.

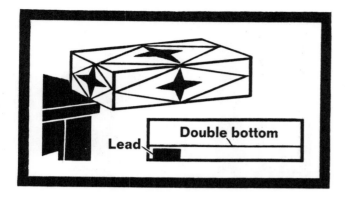

Gravity
Bewitched box

Stick a false bottom in a thin cardboard box and hide a lead weight in the space below. You can always balance the box on the corner in which the piece of lead is lying.

Each object has a centre of gravity, round which it is held in balance by the force of gravity. In such a regularly shaped object as a box the centre of gravity is exactly in the middle, so your box should fall from the table. The lead weight prevents this, however, by shifting the centre of gravity above the table.

The balancing button

If you place a button on a cup so that only the edges are in contact, it will fall off at once. No-one would think that the button would remain on the rim of the cup if you fixed yet another weight on to it. And yet it is possible. If you fix two kitchen forks over the button and then place it on the rim of the cup, it will remain in this position.

The bent fork handles, whose ends are particularly heavy and reach sideways round the cup, move the centre of gravity of the button exactly over the rim of the cup, so that the whole set-up is in balance.

129

Floating beam

It would not seem possible to balance a clothes peg with one end on the tip of your finger if a leather belt is hung over half the peg. But the force of gravity can apparently be overcome.

The whole secret is a small nick which you cut slantwise in the piece of wood. The belt, which you squeeze firmly into the nick, leans so far sideways because of its slanting fixing that the centre of gravity of wood and belt together is shifted under the tip of the finger and balance is obtained.

130

Candle see-saw

Push a darning needle sideways through a cork and fix equal-sized candles at both ends. Then push a knitting needle lengthways through the cork and lay it over two glasses. If you light the candles, they begin to swing.

Before the candles are lit, the centre of gravity of the see-saw lies exactly on the axis so that both ends are balanced. But as soon as a drop of wax falls at one end, the centre of gravity shifts to the other side. This is now heavier and swings down. Since the candles drip alternately, the centre of gravity moves from one side to the other.

Letter balance

Stick a penny or a half-dollar (or a metal disk of similar weight) at the top right-hand corner of a picture postcard and fix two paper clips on the opposite corner. Hang the card up on a wooden surface with an easily-turned pin in the top left-hand corner. The most simple of letter balances is complete and with it you can check the weight of a letter as accurately as with a normal type letter balance.

You must, however, first standardize your balance. Hang a letter which weighs exactly four ounces onto the paper clips and mark the displacement of the top right-hand corner by an arrow on the wall. With letters of more than four ounces the balance moves further and you know that you need more postage. This simple construction is a first-order lever, which is suspended by a pivot just like a normal letter balance. The left-hand narrow edge of the card forms the loading arm, and the upper edge the force arm, which shows, because it is longer, even small differences in weight.

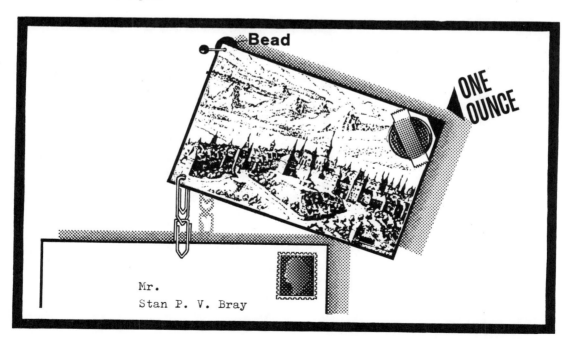

Bead

ONE OUNCE

Mr.
Stan P. V. Bray

132

Magic rod

Lay a rod over your index fingers so that one end sticks out further than the other. Will the longer end become unbalanced if you move your finger further towards the middle?

The rod remains balanced however much you move your finger. If one end becomes over-weight it presses more strongly on the finger concerned. The less loaded finger can now move further along until the balance is restored. The process can be repeated by the combined effects of the force of gravity and friction until the fingers are exactly under the centre of the rod.

133

Boomerang tin

Make slits half an inch wide in the middle of the bottom and lid of a round biscuit tin. Push a piece of thick rubber the same length as the tin through the slits, and tighten it from the outside with pins. Hang a nut of about two ounces on to the centre of the rubber with a paper clip. If you roll the tin several yards forwards, it will return at once.

The force of gravity prevents the nut from joining in the rolling movement of the tin. It hangs upright under the rubber and winds it up at each rotation. A force is produced in the rubber by the tension, and this causes the backward movement.

Balancing acrobat

Trace the clown below on to writing paper, cut out two figures and glue both pieces together. Stick two small coins into the hands so that they are invisible, and colour the figure brightly. The little paper clown will balance everywhere, on a pencil point, on your finger or as a tight-rope walker on a thread. This trick baffles everybody. It would seem that the figure should fall because its top half is apparently heavier.

The weight of the coins cause the centre of gravity of the figure to shift under the nose, so that it remains balanced.

To be traced

Interplay
of forces
Paper bridge

Lay a sheet of writing paper as a bridge across two tumblers, and place a third tumbler on it. The bridge collapses. But if you lay the paper in folds, it supports the weight of the tumbler.

Vertical surfaces are much less sensitive to pressure and stress than those laid flat. The load of the tumbler is distributed over several sloping paper walls. They are supported in the folds and thus have a very high stability. In practice the stability is increased by moulding sheets and slabs to give rounded or angled sections. Now think about corrugated iron and corrugated paper.

Unbreakable box

Put the outside case of a match box on the table, and on its striking surface place the box itself. Probably everybody would suppose that the match box could be smashed with one blow of the fist. Try it! The box nearly always flies off undamaged in a high curve.

The match box is so strong because of its vertically joined sides that the pressure of the striking fist is transmitted to the outside without smashing it. The case, whose side walls seldom stand absolutely vertical, diverts the pressure to the side.

137

Strong egg

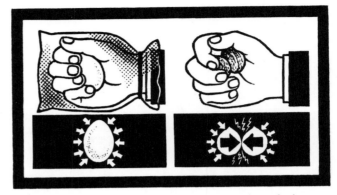

Who wants to bet that you can crack walnuts more easily in your hand than a raw hen's egg? Take a slightly polished egg, place your hand as a precaution in a plastic bag and press as hard as you can!

The lever pressure of the fingers is distributed evenly from all sides on to the egg and is not enough — if the shell is undamaged — to break it. Curved surfaces are extremely stable. People use this advantage in the building of arches and bridges, and cars hardly have a flat surface for the same reason. Two walnuts can be easily cracked in one hand, because the pressure is concentrated at the points of contact.

138

Knot in a cigarette

You would not think it was possible to tie a knot in a cigarette. Wrap the cigarette in cellophane and twist up the ends firmly. It is now possible, without more trouble, to make a knot in the cigarette without breaking it.

The cigarette would split at once without the cellophane casing, because on bending the pressure of the tobacco filling would push through the paper at the point of greatest strain. The cellophane casing is so tough that it distributes the pressure from the inside evenly over the whole length of the cigarette. After untying the knot and unwrapping, you only need to smooth out the cigarette.

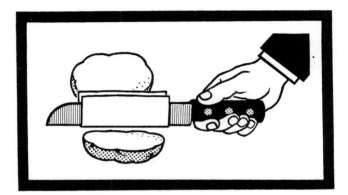

Uncuttable paper

Place a folded piece of writing paper round a knife blade. You can cut potatoes with it without damaging the paper.
The paper is forced into the potato with the knife. It is not cut itself because the pressure of the blade on the paper meets a resistance from the potato. Since its flesh is softer than the paper fibre, it yields. If, however, you hold the paper firmly on top, the pressure balance is lost, and the paper is broken.

Spinning ball

Place a marble on the table, with a jam jar upside down over it. You can carry the ball in the jar as far as you like, without turning it the right way up. Anybody who does not think it is possible should make a bet with you.
It is made possible by a little physical trick: make turning movements with the jar and thus set the ball rotating too. The ball is pressed against the inner wall of the jar by the centrifugal force. The narrowing of the glass vessel at its mouth stops the ball flying out when you lift the jar from the table.

141

Egg-top

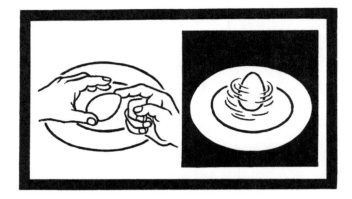

There is a very simple method for distinguishing a cooked egg from a raw one without breaking the shell. Spin the eggs on a plate, and the well-cooked one continues to rotate. Since its centre of gravity lies in the thicker half, it even stays upright, like a top.

The liquid inside the raw egg prevents this. Since the yolk is heavier than the white, it rolls from the middle when you spin the egg because of the centrifugal force. It brakes the movement so much that it only amounts to a clumsy rocking.

142

Coin bumping

Lay three coins in a row on the table so that two touch. Press the middle one hard with your thumb and flip the coin lying apart against it. The neighbouring coin shoots away, although the middle one is held firm.

Solid bodies possess a more or less large elasticity, that is shown, for example, in steel when it is made into a spring. In our experiment the coins are imperceptibly compressed when they collide but spring back at once to their original shape and transfer the impact to the neighbouring coin in this way.

Dividing money

Lay several coins of the same value in a row on an even surface so that they touch. Lay another coin a short distance away from the row, and flip it with your finger against the others. What happens? A coin slides away at the other end of the row. Repeat the experiment, pushing two coins against the row. This time two coins separate off. If you flip three coins, three are removed, and so on. In a further experiment you can push the coins really hard, but with no different result.

The result is truly surprising, but illustrates a physical law. The coins undergo an elastic impact when they are knocked together by which the same weight as the flipped coins carries on the movement at the other end of the row. The sharpness of the flip decides how fast and how far the coins fly off, but has no effect on their number.

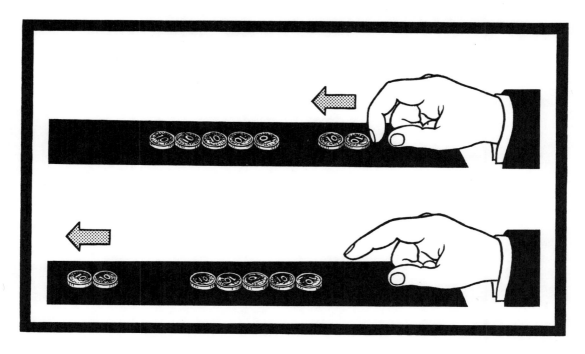

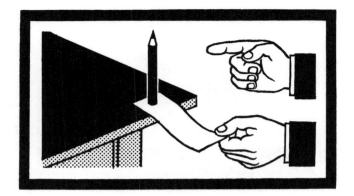

Inertia
The stable pencil

Hold a strip of paper over a smooth table edge and place a pencil on it. Can you remove the paper without touching the pencil or knocking it over? The pencil will certainly fall if you pull the paper away slowly. The experiment works if you take the paper away in an instant by hitting it with your finger.
Each body tries to remain in the position or state of motion in which it finds itself. The pencil resists the rapid movement, so that it remains where it is and does not tip over.

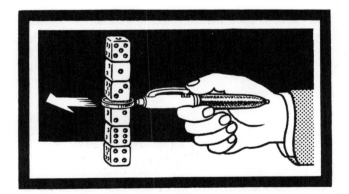

Treasure in the tower

Build six dice made of ivory or plastic into a tower and place a coin in the middle, using a quarter-dollar or a 5 new-penny piece. The tower is very rickety: then how can you remove the coin without touching it or knocking it over?
Hold a ball-point pen with a pushing device a little way from the coin. If you discharge it the coin flies out of the tower. The movement of the spiral spring in the ball-point pen is transferred at once to the coin, but because of the low friction not to the dice, which because of their weight have a fairly large inertia.

146

Egg bomb

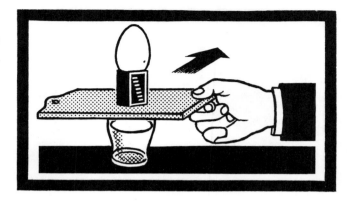

Lay a small board on a glass of water, place on it a match box case, and on this a raw egg. Can you transfer the egg to the water without touching it? Pull the board sharply to the side! The egg falls undamaged into the water.

The inertia of the egg is so great because of its weight that it is not carried along with the fast movement. The light box, on the other hand, flies off because its inertia is low.

147

The lazy log

Tie two pieces of cord of equal thickness to a block of wood or another heavy object. Hang the wood up by one cord and pull on the other. Which section of cord will break?

If you pull slowly, the strain and the additional weight of the object causes the upper cord to break. But if you pull jerkily, the inertia of the block prevents the transfer of the total force to the upper cord, and the lower one breaks.

Cut far enough into the flesh of an apple with a knife, so that when you lift it up it sticks to the blade. Now knock against the blade in the apple with the back of another knife. After several blows the apple will break in half.

The famous Italian scientist Galileo discovered in the sixteenth century that all bodies offer a resistance to an alteration in their position or state of motion, which we call inertia. This prevents the apple from following the jerky movements of the knife. It pushes slowly on the blade, until it is separated.

Place about 20 coins of the same value in a pile on a smooth table. How can you take away the coins one at a time from below, without touching them? Flip another coin sharply with your finger so that it hits the bottom coin and shoots it away.

If you aim well, you can shoot away all the coins in this way. The inertia of the coin column is so great that the force of the flipped coin is not sufficient to move it or completely overturn it.

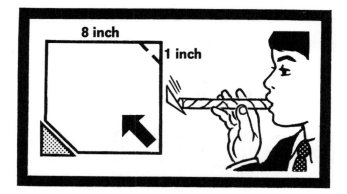

Sound
Humming flute

A square piece of paper has one corner snipped off, and two notches made in the opposite corner. Roll the paper in the direction of the arrow in the figure to make a tube about as thick as a pencil and push the notched corner back into the opening. Draw a deep breath through the tube. This causes a loud humming note.
The paper corner is sucked up by the air which is drawn in, but since it is slightly springy, it begins to vibrate. The vibration is quite slow, so the note is deep.

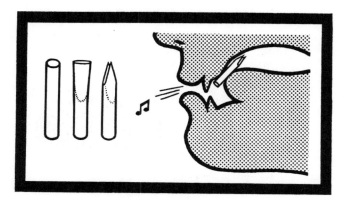

151

Musical drinking straw

Cut a piece about an inch long from a plastic drinking straw. Press one end together and cut it to a point. If you fix the straw against the front of the top palate, you can obtain various notes when you blow through it.
The pointed tongues of the straw are moved in very rapid sequence by the air sweeping through, so that the note produced is fairly high. A large number of musical instruments are based on this simple principle.

152

Water organ

Half fill a thin walled glass with water, dip in your forefinger and run it slowly round the rim of the glass. A lovely, continuous ringing note is produced.
The experiment only works if your finger has just been washed. It rubs over the glass, giving it tiny jolts. The glass begins to vibrate, which produces the note. If your finger is greasy, it slides smoothly over the glass without the necessary friction. The pitch of the note depends on the amount of water, and the vibration can be seen clearly on the water surface.

153

Note transfer

You can extend the previous experiment. Place two similar thin walled glasses an inch apart and pass your freshly washed finger slowly round the rim of one of them. A loud humming note is produced. In a strange way the second glass vibrates with the first, and you can observe this vibration if you place a thin wire on it.
The vibration of the first glass is transmitted to the second by the sound waves in the air. This 'resonance' only occurs if the glasses produce notes of the same pitch when struck. If this is not the case with your two glasses, you must pour some water into one until the pitches are the same.

154

Peal of bells

Tie a fork in the middle of a piece of string about a yard long. Wind the ends several times round your forefingers and hold the tips of your fingers in your ears. Let the fork strike a hard object. If the string is then stretched, you will hear a loud, bell-like peal.

The metal vibrates like a tuning fork when it strikes the hard object. The vibration is not carried through the air in this case, but through the string, and the finger conducts it directly to the eardrum.

155

Paper diaphragm

Halve a match, make a point on it and split it at the other end. Fix it on to a smooth piece of paper and hold it vertical on an old turning phonograph record. The music sounds over the paper almost as clearly as from a loudspeaker.

The match travels in the grooves of the record and transfers the lateral vibrations to the paper like a phonograph needle to the diaphragm of a loudspeaker. The vibrations of the paper are carried to your ear drum as sound waves through the air.

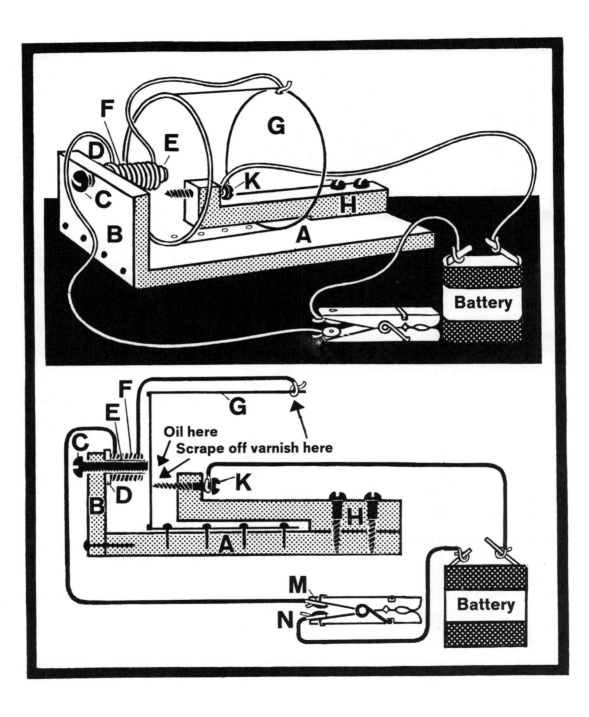

Oil here
Scrape off varnish here

Box horn

Nail as large a tin can as possible firmly onto a suitable board A. Through board B, which is fixed at the side, bore a hole through which you can turn an iron screw C to the middle of the base of the tin. A small space should remain between screw and tin. Put a layer of paper E round the screw and over it wind about two yards of covered wire F. From the inside a wood screw K, which is fixed so it can be moved in a piece of wood H, contacts the base of the tin. Scrape off the metal plating in front of the tip of the screw and oil it. Join all the parts correctly with connecting wire, removing the insulation and tin varnish from the connecting points. A clothes peg with two drawing pins M and N acts as the horn button. If you press it, you will obtain a very loud signal.

The apparatus works on the same principle as a car horn. If you close the circuit by pressing the horn button, the screw C becomes magnetic and attracts the base of the tin. So the circuit is broken in front of the screw K. Screw C loses its magnetism, and the base of the tin springs back to the screw K. The process is repeated so quickly that the tin plate produces the horn note by its vibration.

Footsteps in a bag

Put a house fly in a smooth paper bag, close it, and hold it horizontally above your ear. If you are in a quiet room you can hear the patter of the six legs and other rather curious noises quite clearly.

The paper behaves like the skin of a drum. Although only the tiny legs of the insect beat on it, it begins to vibrate and transmits such a loud noise that you would imagine that a larger organism or a rattling clockwork motor was in the bag.

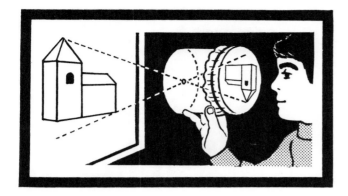

158
Light
Pinhole camera

Bore a hole in the middle of the base of a box. Stretch parchment paper over the mouth of the box and secure it with a rubber band. If you focus this simple camera on a brightly lit building from a dark room, the image appears upside down on the screen.

Our eyes work on the same principle. The light rays fall through the pupil and lens and project an inverted image on the retina. The image is turned the right way up again in the sight centre of the brain.

159
Drop microscope

Bore a hole about one fifth of an inch wide in a strip of metal and smooth the edges. Bend the metal so that you can fix it with adhesive tape half an inch above the bottom of a thin glass. A pocket mirror is placed inside on a cork, so that it is on a slant. If you dab a drop of water into the hole, you can see small living organisms and other things through it, magnified by up to fifty times.

The drop magnifies like a convex lens. When you bring your eye near to it the sharpness can be adjusted by bending the metal inwards. The angle of the mirror is adjusted automatically by moving the glass.

160

Fire through ice

You would hardly believe it, but you can light a fire with ice! Pour some water which you have previously boiled for several minutes into a symmetrically curved bowl, and freeze it. You can remove the ice by heating slightly. You can concentrate the sun's rays with the ice as you would with a magnifying glass and set thin black paper alight, for instance.

The air in fresh water forms tiny bubbles on freezing and makes the ice cloudy. But boiled water contains hardly any air and freezes to give clear ice. The sun's rays are only cooled imperceptibly when they pass through the ice.

161

Shortened spoon

Look from just above the rim of a bucket of water, and dip a spoon upright into it. The spoon seems to be considerably shorter under the water.

This illusion is based on the fact that the light rays reflected from the immersed spoon do not travel in a straight line to your eyes. They are bent at an angle at the surface of the water, so that you see the end of the spoon higher up. Water always seems more shallow than it actually is because of the refraction of light. The American Indians also knew this. If they wanted to hit a fish with an arrow or spear, they had to aim a good deal deeper than the spot where the fish appeared to be.

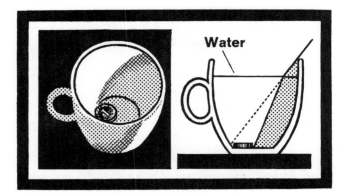

Lay a penny in a cup near the side. Place the cup in oblique light so that the shadow of the rim just covers the coin. How can you free the penny from the shadow without moving the cup or the coin or using a pocket mirror?

Quite simple! Bend the light rays back to the coin. Fill the cup with water and the shadow moves to the side. The light rays do not go on in a straight line after striking the surface of the water, but are bent downwards at an angle.

163

Broken pencil

Half fill a glass with concentrated salt solution and top it up with pure water using a spoon. If you hold a pencil at the side of the glass, it seems to be broken into three pieces.

The light rays coming from the immersed pencil are bent at an angle when they emerge from the water into the air at the side of the glass. Because salt solution has a different composition from pure water, the angle of refraction is different. We know that how much light rays are bent when they pass from one substance into another entirely depends on the 'optical density' of the different substance.

164

Cloud of gas

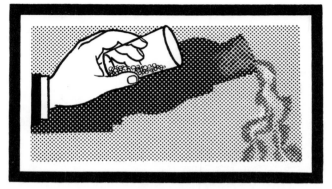

If you pour some bicarbonate of soda and vinegar into a beaker, carbon dioxide is given off. You can normally not see the gas, but it can be made visible: tilt the beaker with its foaming contents in front of a light background in sunlight. You can see the gas, which is heavier than air, flowing from the beaker in dark and light clouds.
Carbon dioxide and air have different optical densities, and so the light rays are bent when they pass through them. The light clouds on the wall are formed where by refraction the propagated light is bent towards it, and the dark clouds are seen where light is bent away.

165

Bewitched pencils

Look through a round jam jar filled with water. If you stand a pencil a foot behind it, its image appears doubled in the jar. If you close your left eye, the right-hand pencil disappears, and if you close your right eye, the other goes.
One sees distant objects reduced in size through a normal magnifying glass. The water container behaves in a similar way, but since it is cylindrical, you can look through it from all directions. In our experiment both eyes view through the jar from a different angle, so that each one sees a smaller image for itself.

Secret of 3-D postcards

Draw red and blue vertical lines a short distance apart and lay over them along their length a round, solid, transparent glass rod. You can see both lines through the glass. But if you close one eye the red line disappears, and if you close the other eye the the blue line disappears.

Each eye looks from a different angle through the rod and perceives — by the particular angle of light refraction — only one line. The experiment explains how the stereoscopic postcard works: its surface consists of thin, transparent ripples, which behave like our glass rod.

Two photographs, each taken from a different angle, are copied together in very fine vertical strips to give a picture, so that under each individual ripple lies a strip of one and a strip of the other photograph. In the ripples we see, as with our glass rod, only the strip of one photograph with each eye, and the brain finally joins the images to give a 3-D picture.

167

Finger heater

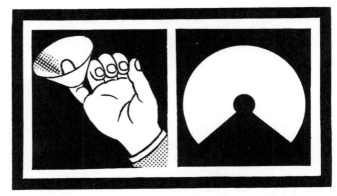

Glue a funnel together with smooth silver paper, as shown in the picture. Stick your finger into it, point it to the midday sun, and you will feel it warm up quite a lot.

The sun's rays are reflected from the walls of the funnel to the middle and are concentrated on the central axis, which is formed by your finger. If you put your finger into the dismantled concave mirror of a bicycle lamp, the sun's rays would be unbearably hot. In this case they converge at a point, the focal point of the concave mirror, at which the bulb is usually placed. The heat produced is so great that one could easily start a fire with a concave mirror.

168

Sun power-station

The sun's radiation can be caught in a bowl and by means of the heat potatoes can be stewed in their own juice. A 'nourishing' joke and an instructive experiment at the same time. Take a soup bowl or a large salad bowl with as small a base as possible and line it inside with household aluminium foil — bright side outwards. Smooth the folds with a rubber ball and a spoon until the foil acts like a mirror. Split it a little at the base of the bowl so as to be able to press in a suction hook, on which you fix a small raw potato. If you point the cooker on a warm day towards the midday sun, the potato becomes hot at once and is cooked after some time.

Now and then you must re-align the bowl towards the sun. The sun's rays falling on the aluminium foil are reflected to the middle and concentrated on the potato. In tropical countries people often use concave mirrors for cooking. Did you know that even electricity can be produced in large power stations by the sun's radiation?

Aluminium foil

Suction hook

Adhesive tape

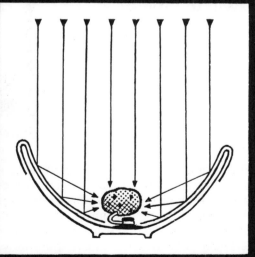

169

Magic glass

If you place a jar over a coin lying on the table, it looks just as if it were in the jar. If you now pour water into the jar and put the lid on it — abracadabra! — the coin has disappeared, as if it had dissolved in the water.

When the jar is empty the light rays from the coin travel into our eyes in the usual way. But if the jar is filled with water, the light rays do not follow this path any more. They are reflected back over the bottom of glass when they hit the water from below at an angle. We call this total reflection, and only a silvery gleam can be seen on the bottom of the jar.

170

View into infinity

Hold a pocket mirror between your eyes so that you can look to both sides into a larger mirror. If you place the mirrors parallel to one another, you will see an unending series of mirrors which stretch into the distance like a glass canal.

Since the glass of the mirror shines with a slightly greenish tint, some light is absorbed at each reflection, so that the image becomes less sharp with increasing distance. Nevertheless the experiment is interesting, because one can make an image of infinity for oneself.

Kaleidoscope

You need a highly glazed picture postcard. Cut the edges smooth and divide the writing side along its length into four panels an inch wide. Scratch the lines lightly and bend and stick the card into a triangular tube-shape with the glazed side facing inwards. Both openings are glued up with transparent cellophane. At one end also stick white paper over the cellophane, having previously inserted small snips of coloured cellophane in between, so that they can move easily. A beautiful pattern, which alters on tapping with your finger, appears in the tube.

The three highly glazed surfaces of the bent picture postcard behave like mirrors and multiply the image of the coloured pieces of cellophane. A polished surface reflects better, the flatter the light rays hit it. But since part of the radiation is absorbed into the surface, the image reflected from it is not so clear and bright as with a mirror.

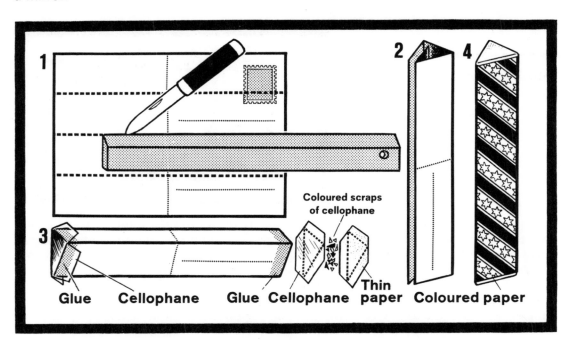

Coloured scraps of cellophane

Glue   Cellophane   Glue Cellophane   Thin paper   Coloured paper

172

Mirror cabinet

Obtain three sections of mirror each about 3 × 4 inches in size, or cut them yourself. Polish them well, and join them with adhesive tape — reflecting surfaces facing inwards — to make a triangular tube. Stick coloured paper outside. If you now look obliquely from above into the mirror prism you will discover a magic world! If you hold a finger in the prism, its image is always multiplied six times in an endless series in all directions. If you place a small flower inside, a meadow of flowers stretches into the distance. And if you move two small figures, innumerable couples dance in an immense hall of mirrors.

173

Shining head

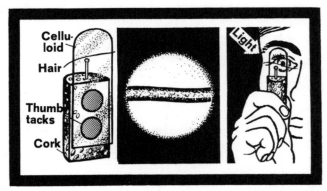

Stick a pin with a polished head into a cork cut in half lengthways and fix some celluloid on it to protect your eyes. If you look at the tiny light reflection from the head of the pin under a bright lamp, while holding it right up to the eye, it appears as a plate-sized circle of light. A hair stuck onto the moistened celluloid is seen magnified to the width of a finger in the circle of light.

The head of the pin behaves like a small convex mirror. The light which hits it is spread out on reflection, and irradiates a correspondingly large field on the retina of the eye.

Light mill

Cut a two-inch circle from heavy aluminum foil. With a pencil point, make a small dent exactly in the middle to act as a pivot. Cut six evenly spaced slits in the aluminum and fold each of the blades perpendicular, with the shiny side of the aluminum always facing the same direction. Cover the matte side of each blade with black tempera paint. Next, make a small hole in the screw-on lid of a marmalade jar. Balance the propellor on a needle, stick the needle into a piece of cork and glue the cork to the inside of the lid. Now cover the screw threads of the jar completely with glue, and screw the jar firmly to the lid. Warm the jar briefly on the stove to make the air inside expand, and cover the hole immediately with tape. When the air cools, there will be a partial vacuum in the jar which reduces the air resistence to the propellor. When you place the radiometer in sunlight or under a bright lamp, the propellor will spin without stopping. – We know that dark-colored surfaces heat up faster in sunlight than light-colored ones. The black sides of the blades absorb more light rays and become ten times warmer than the silver sides. The difference in the amount of heat radiated by the two sides of the blades causes the rotation.

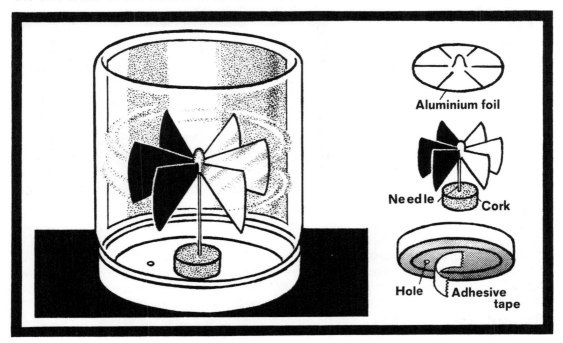

Aluminium foil

Needle   Cork

Hole   Adhesive tape

175

The sun's spectrum

Lay a piece of white paper on the window sill and place on it a polished whisky glass full to the brim with water. Fix a postcard with a finger-wide slit onto the glass, so that a band of sunlight falls onto the surface of the water. A splendid spectrum appears on the paper, and the bands red, orange, yellow, green, blue, indigo and violet can be easily distinguished.

The experiment is only possible in the morning or evening when the sunlight falls obliquely. It is refracted at the surface of the water and the side of the glass and is separated as well into its coloured components. The experiment also works well with light from an electric torch.

176

Spectrum in a feather

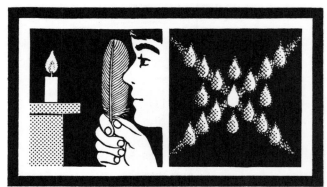

Hold a large bird's feather just in front of one eye and look at a burning candle standing a yard away. The flame seems to be multiplied in an X-shaped arrangement, and also shimmers in the spectral colours.

The appearance is produced by the bending of light at the slits. Between the regular arrangement of feather sections (vanes and barbs) are narrow slits with sharp edges. The light is bent on passing through them, that is, it is refracted and separated into the spectral colours. Since you see through several slits at the same time, the flame appears many times.

You will certainly only have seen a rainbow in the sky as a semi-circle up to now. You can conjure up a complete circle for yourself from sunlight. Stand out-of-doors on a stool in the late afternoon with your back to the sun and spray a fine shower with the water hose. A coloured circle appears in front of you!

The sunlight is reflected in the drops so that each shines with the spectral colours. But the colours of the drops are only visible to your eyes when they fall in a circular zone at a viewing angle of 85° in front of you. Only the shadow of your body briefly breaks the circle.

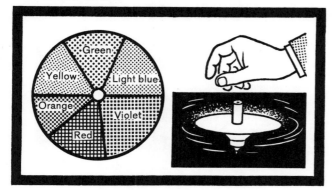

Cut a circle about four inches in diameter from white cardboard and colour it as shown with bright-coloured felt pens. Stick the disk on a halved cotton reel, push a pencil stump through it and allow it to spin. The colours disappear as if by magic, and the disk appears white.

The colours on the disk correspond to the colours of the spectrum of which sunlight is composed. On rotation our eyes perceive the individual colours for a very short time. However, since the eyes are too sluggish to distinguish between the rapidly changing colour impressions, they merge and are transmitted to the brain as white.

179

Summer lightning
(see back cover)

Look alternately left and right at the blue of the sky. You will not trust your eyes because it flashes continuously with bright lightning.

What is the explanation for this appearance? If you look at the picture, it is imprinted on the retina of the eye. But red colour impressions remain longer on your retina than blue when you move your gaze. So the impression of the red lightning is overlaid for an instant on the blue of the sky. These two colours together, however, produce an impression of bright light in your brain. Since a new impression of the lightning is formed with each movement of the eye, the process is repeated.

180

Unusual magnification

Make a small hole in a card with a needle. Hold it close to the eye and look through it. If you bring a newspaper very close you will see, to your surprise, the type much larger and clearer.

This phenomenon is caused in the first place by the refraction of light. The light rays passing through the small hole are made to spread out, and so the letters appear larger. The sharpness of the image is caused – as in a camera – by the shuttering effect of the small opening. The part of the light radiation which would make the image blurred is held back.

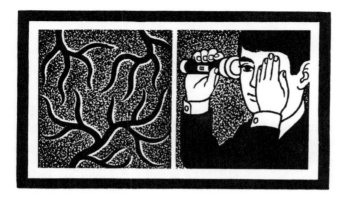

Veined figure

Close your left eye in a dark room and hold a lighted torch close beside the right eye. Now look straight ahead and move the torch slowly to your forehead and back. After some time you will see a large, treelike branched image in front of you.
Very fine blood vessels lie over the retina of the eye, but we do not normally see them. If they are irradiated from the side, they throw shadows on the optic nerves lying below and give the impression of an image apparently floating in front of you.

Motes in the eye

Make a hole in a card with a needle and look through it at a burning, low-power electric light bulb. You will see peculiar shapes which float before you like tiny bubbles.
This is no optical illusion! The shapes are tiny cloudings in the eyes, which throw shadows on to the retina. Since these are heavier than the liquid in the eye, they always fall further down after each blink. If you lay your head on one side, the motes struggle towards the angle of the eye, showing that they follow the force of gravity.

183

Ghost in the castle

A white ghost haunts these ruins at night! Would you like to see it? Hold the picture with the black figure upright at the normal reading distance in front of your eyes and look fixedly at its mouth in bright light for one minute. If you then look at once in the tower of the ruined castle, there will appear after about 10 seconds the illusion of a white ghost.

When you look at the figure, part of the retina is not exposed to light from the black surface. The rest of the optic nerves are dazzled by the bright paper and tire quickly. If you now look at the castle tower, the tired optic nerves do not see the white of the paper in its full brightness, but as a grey surface. The rest, on the other hand, see the white tint of the paper all the more clearly. So an exchange of the black and white surfaces is produced and you see a white ghost in the dark arch of the tower. Only after quite a time, when the nerves have adjusted themselves, does the ghost disappear.

Goldfish in the bowl
(see back cover)

Stare in bright light for one minute at the eye of the white fish. If you then look at the point in the empty gold fish bowl, there appears to be light green water and a red fish in it after several seconds.

If the eyes have stared for a long time at the left-hand picture, the part of the light-sensitive retina which is irradiated by the red surface tires and the optic nerves concerned become rather insensitive to red. So on looking at the white surface in the right-hand picture, they do not perceive the red radiation which is present in white light. They are only sensitive to the yellow and blue components, which together give green. But the part of the retina which has received the picture of the white fish is now sensitive to the opposite colour to green, namely red. Coloured after-images can be produced with other colours just as well. Each colour changes into the opposite: i.e. blue into yellow, yellow into blue and green into red.

Bewitched rabbit

Look at the picture at the normal reading distance. Then shut your left eye and stare at the magic rod with your right. If you now slowly alter the distance of the picture — abracadabra — the rabbit suddenly disappears.

The retina of the eye consists of a large number of light sensitive nerve endings, the so-called rods and cones. These are absent at one point, where they join together at the optic nerve. If the image of the rabbit thrown on the retina falls at this 'blind spot' as we move the picture we cannot see it.

186

The disappearing finger

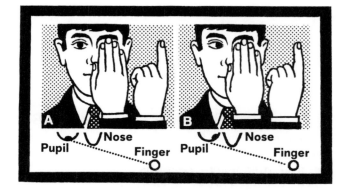

Cover your left eye with your right hand and look straight ahead with your right eye. Raise your left forefinger to your left ear and move it until the tip of the finger is just visible (A). If you now move your eye to look directly at the finger (B), strangely enough it disappears.

This interesting experiment has a geometrical explanation: when you are looking straight ahead (A) the light rays from the finger pass over the bridge of your nose into the pupil of the eye. But if the pupil is moved to the left (B) the light rays from the finger go past it.

187

Hole in the hand

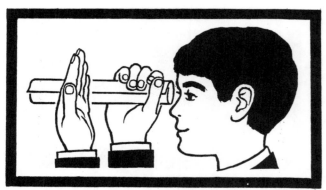

Roll a piece of writing paper into a tube and look through it with your right eye. Hold your left hand open on the left next to the paper. To your surprise you will discover a hole, which apparently goes through the middle of the palm of your hand. Can you think what causes this illusion?

The right eye sees the inside of the tube and the left the open hand. As in normal vision, the impressions which are received by each eye are combined to give a composite image in the brain. It works particularly well because the image from inside the tube, which is transferred to the palm of the hand, is in perspective.

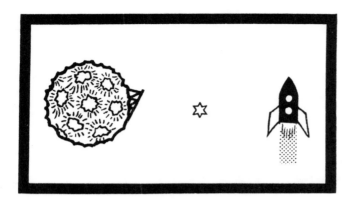

Hold the picture so that the tip of your nose touches the star, and turn it round slowly to the left. The rocket flies into the sky and lands again on the moon. Each eye receives its own image on viewing and both impressions are transmitted to give a composite whole in the brain. If you hold the star to the tip of your nose, your right eye only sees the rocket and the left only the moon. As usual, the halves of the image are combined in the brain. As you turn the picture on its edge, it does not shrink any more because both eyes see the same image by squinting.

Hold your forefingers so that they are touching about a foot in front of the tip of your nose and look over the fingertips away to the opposite wall. On doing this you will see a curious ball, which is apparently fixed between the fingertips.

When you look over your fingers your eyes are focussed sharply on the wall. But the fingers are then projected on the retina in such a way that the images are not combined in your brain. You see the tips of both fingers doubled. These finally combine to give the illusion of a round or oval image.

Illusions
Two tips to your nose

Cross over the index and middle fingers and rub them sideways over the tip of your nose. To your great surprise you will feel two noses.
When you cross them over, the position of the sides of the fingers is exchanged. The sides normally facing away from one another are now adjacent, and both touch the tip of the nose together. Each one reports separately, as usual, the contact with the nose to the brain. This is deceiving because the brain does not realize that the fingers have been crossed over.

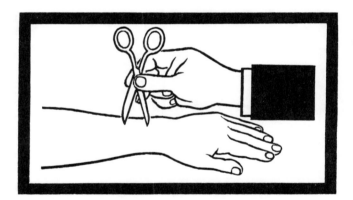

Touch test

To test a friend's sense of touch, ask him to close his eyes. Open a pair of scissors about an inch wide and touch his arms with both points at once. Your friend will only feel one point. Repeat the experiment on other parts of the body.
The experiment shows that man's sense of touch is differently developed on different parts of the body. On the back the sensitive nerve ends are not so abundant as, for example, on the face. On the hand and especially at the fingertips there are so many of them that one would feel both points of even a slightly opened pair of scissors.

192

Reaction time

Hold a pencil above your friend's slightly open fist and ask him to catch the falling pencil by closing his hand. He cannot do it!
When the eyes see the pencil fall they first send a signal to the brain, and from here the command 'Grab' is sent to the hand. Time is naturally lost in doing this. If you try the experiment yourself, it must succeed, because the commands to let fall and to grab are simultaneous. We call the time between recognition and response, the reaction time. The time lost in a dangerous situation can mean death for a car driver.

193

Confused writing

Do you bet that you cannot write your name if you make circular movements with your leg at the same time? You will manage nothing more than an unreadable scribble.
It is probably possible to draw in the same direction as the circular leg movements. But as soon as you circle your leg in the other direction, the pencil movements cross over completely. So the leg movements are transferred to the writing. Each action needs so much concentration that both cannot be carried out at the same time. Your concentration is broken in a similar way if you do school work while you are listening to music.

Hold a card in front of your forehead and try to write your name. You will be surprised at what appears. Your name is in mirror writing.

From pure habit you have started at the left and finished at the right, as you usually do when writing. This was a mistake! Because if you had thought about it, the writing must be laterally reversed.

Put a bottle upright on the ground and walk round it three times. If you then try to walk straight towards something, you will not be able to do it.

The balancing organ in the inner ear has played a trick on you. A liquid starts to move in it when you turn your head. Small hairs are bent when this happens and report the process to the brain. This makes sure that you make suitable counter-movements. But if you turn quite quickly and stop suddenly, the liquid goes on moving. The brain reacts as if you were still turning and you go in a curve.

196

Annoying circles

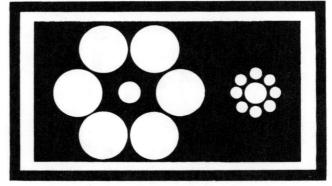

Look at these two figures. Which of the two circles in the middle of the figures is larger?
Both circles are the same size! In the subconscious mind we do not only compare the middle circles with one another, but also with those in the surrounding circles. In this way we get the impression that the middle circle in the right-hand figure is larger. We are subject to a similar optical illusion when we look at the moon. If it is close to the horizon, we compare it with houses and trees. It then appears larger than when it is high in the sky.

197

Magic spiral

Look at the picture closely. You will probably be certain that it is of a spiral.
But a check with a pair of compasses shows that the picture is of concentric circles. The individual sections of the circles apparently move spirally to the middle of the picture because of the special type of background.

Make a point on a piece of paper and place it in front of you on the table. Now try to hit the point with a pencil held in your hand. You will manage it easily. But if you close one eye, you will almost always miss your target.

The distance can only be estimated with difficulty with one eye, because one normally sees a composite image with both eyes and so can discern the depth of a space. Each eye stares at the point from a different angle (notice how the angle alters if you move nearer to the point). The brain can then determine the distance of the point fairly accurately from the size of the angle.

In the picture, letters are embroidered on check material with yarn which is twisted from a black and white thread. Do you have any doubt that the letters are sloping? A ruler will show that the letters are straight. Because of the sloping bars in the background and the twisted threads, our eyes experience a confusing shift in the outlines of the letters.

200

The turning phonograph

If you move the phonograph in front of your eyes in a circle the record appears to turn.

This movement of the record has many causes. When we move the picture, the constantly changing light incidence and angle of observation produce in the eye moving dark and bright zones, which apparently move across the record. The eye cannot follow so quickly, and sees normal rotation of the record.

201

Money fraud

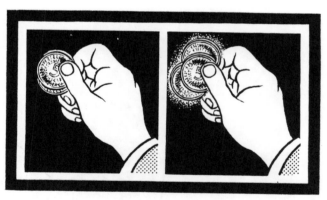

Hold two coins of the same value between thumb and forefinger, and rub them together quite quickly. If you look carefully, you will detect yet a third coin, which apparently moves backwards and forwards between the others! What causes this startling illusion?

Our eyes react too slowly to follow the rapid to-and-fro movement of the coins. Each time the image of the coins remains for a little while on the retina, although they have already moved away. So we see both coins in movement and the after-image of a third coin.

Rocking pudding
(see back cover)

Hold the coloured picture with the pudding – preferably in dim lamplight – at the normal reading distance from the eyes, and move it to and fro sideways. The pudding seems to rock and almost spills over the dish.

The rocking occurs because the impression of warm colours (e.g. red and brown) stays longer on the retina of our eyes than cold colours (such as blue and green). When you move the picture, the background and dish move together in the normal way. The pudding follows the movement, but a little while later. A real pudding behaves quite similarly when it is moved to and fro, because of its inertia, giving this surprising effect.

Living pictures

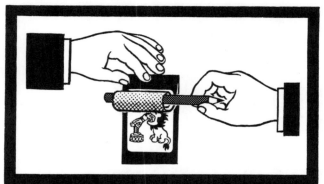

Cut out the two film strips at the top of the page overleaf and stick them – part 1 on part 2 – together at the upper edge. Roll up the top sheet and move it up and down with a pencil. You get the impression that the figure is moving.

The impressions of the pictures received by the eye merge in the brain and produce the effect of movement. This 'cinematographic effect' seems very primitive in this case because it is only produced by two pictures. In normal films 24 pictures roll in a second, and in television as many as 25, and we see smooth and flicker-free movement.

Movies in a cigarette box

Stick one of the film strips from the next page on to drawing paper, separate the eight little pictures, and place them in the right order into a notched cork disk, $\frac{5}{8}$ inch thick. Cut along the side wall of a folding cigarette box for $1\frac{1}{4}$ inches and stick two-thirds of it at right angles facing inwards. Colour the inside of the box black and make holes in the middle of the base and lid. Bend the crank from a worn out ballpoint pen shaft. Fix it into the box and place the cork, which you have previously bored, firmly on it. If you turn the crank round to the right, the figure moves.

Each little picture is shown to the eye for a moment when you wind the crank, and is replaced by another at once. Because the eye is sluggish, each picture leaves an after-image when it has already moved away. The individual pictures merge into one another to give the appearance of movement when you wind the crank. This discovery was made in the year 1830, and today the most modern equipment in the cinema works on the same principle as your table movie.

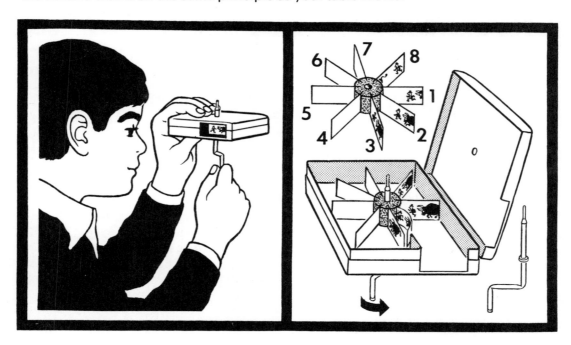

Glue